Distributing Seeds and Tools in Emergencies

Douglas Johnson

Amrita Services
16 Shaftesbury Rd.
Brighton, E. Sussex
BN1 4NE Email Doug@amrita.co.uk

Oxfam

First published by Oxfam GB 1998

A catalogue record for this book is available from the British Library.

ISBN 0 85598 383 3

Available from the following agents:
for Canada and the USA: Humanities Press International, 165 First Avenue, Atlantic Highlands, New Jersey NJ 07716-1289, USA; tel. 732 872 1441; fax 732 872 0717
for Southern Africa: David Philip Publishers, PO Box 23408, Claremont, Cape Town 7735, South Africa; tel. (021) 644136; fax (021) 643358
for Australia: Bush Books, PO Box 1370, Gosford South, NSW 2250, Australia; tel. (043) 233274; fax (029) 212248
for the rest of the world, contact: Oxfam Publishing, 274 Banbury Road, Oxford OX2 7DZ

Published by Oxfam GB, 274 Banbury Road, Oxford OX2 7DZ

JB103/RB/97 Printed by Oxfam Print Unit

Oxfam GB is a registered charity no. 202918, and is a member of Oxfam International.

Contents

Preface

There are very few books available on the distribution of seeds and tools as part of a relief project. There are helpful books on many related subjects, such as assessments, project management, agriculture, evaluation, distribution, gender awareness, food security, working in areas of conflict, extension work, or environmental awareness, but few if any books that specifically cover seeds and tools projects in relief situations. Considering the large number of such projects, especially in Africa, and the considerable sums of money used by organisations to provide seeds and tools, one would expect there to be more extensive literature on the subject.

Oxfam (UK and Ireland) staff involved in seeds and tools projects in various capacities have been concerned about this lack of information for some time. Nigel Taylor, Jane Powell, Michelle Barron, Ros David, and Allyson Thirkell have all been responsible for initiating or conducting preparatory research work, which has been drawn on in writing this book. Their recognition that a practical guidebook on the subject was urgently needed gave the necessary impetus for Oxfam to commission this book. In response to these individual and collective initiatives, Oxfam has produced this straightforward, practical guide, based on field experience, lessons learned, technical data from reliable sources, individual expertise, established practices, and common sense, in the hope that it will be of use to practitioners and planners.

In many aspects of this work, there are no 'right' or 'wrong' ways of doing things, but simply more or less appropriate responses to a specific situation. This is what makes the work so interesting and challenging. In order to produce a guide that is as comprehensive as possible, some material has been included which is common knowledge to many field workers. However, the intention is not only to provide information and ideas, but an overview of the complex process of project development, and a sequential checklist of essential steps.

While this book covers seeds and tools projects mainly from an emergency perspective, it also emphasises the need for long-term thinking and planning. These projects most commonly fit into the category of rehabilitation work, and are often initiated by emergency teams. However, development-oriented teams may well continue to implement them beyond the period of immediate crisis. We hope that this guide will therefore be of use both to those specialising in emergency response work, and to those whose focus is on long-term development work.

The book begins with a consideration of some general development issues relevant to the distribution of seeds and tools. Chapter 2 examines the concept of the project cycle and gives examples of seeds and tools projects and how emergency interventions fit into a long-term framework. The next three chapters deal with information requirements. Chapter 3 covers four broad factors to be considered in the initial 'scoping' stage; Chapter 4 describes the various types of information which will be needed, and Chapter 5 describes in detail how to carry out an initial assessment. The next chapter deals with project design and the process of drawing up a project plan, using the information which has been collected and analysed by the assessment team. Chapter 7 provides technical information on seeds and tools. Chapter 8 describes the activities involved in implementing a seeds and tools project, and in the final chapter the relationships between these interventions and longer-term development work is explored.

1 | General issues related to seeds and tools projects

In this chapter, we will look at some general development issues in relation to projects involving the distribution of seeds and tools: the relationship between different categories of aid work; poverty and livelihoods; the problems of working in situations of conflict and insecurity; what is meant by food security; and how national and international economic policies affect the context of a seeds and tools project.

Emergencies, rehabilitation, and development

How these three separate categories of aid work should be defined, and the relationship between them, have been the subjects of much discussion among aid organisations and development academics in recent years. The apparently very different approaches to each category of intervention taken by different departments within organisations, and between organisations specialising in one of these important areas of work, have been responsible for some antagonism and misunderstandings. After what might be seen as a healthy identity crisis, emergency, rehabilitation, and development work have come to be seen as three phases of one continuum, and three situation-specific responses to fluctuating conditions. It is now generally recognised that there are no clear boundaries between these categories of work.

Increasingly, most aid organisations knowingly move from emergency response slowly into development work; and they knowingly carry out development work where there is a high probability of a crisis occurring. In such situations, planning and preparing for one while doing the other simply makes good sense. Laying the ground for rehabilitation and development phases while implementing emergency work is possible and very prudent. Emergency-preparedness planning and related preparation work makes perfect sense in a situation that is becoming unstable. Projects that directly or indirectly, immediately or in the long term, strengthen the capacity of communities and groups to cope with potential crisis, are the best solution at any stage. Ideally, the aim should be to minimise, if not prevent, crisis; which represents an immense challenge to aid organisations. To begin to achieve this goal will require a thorough understanding of the natural and social cycles and patterns that lead to crisis.

In most cases, projects that are primarily concerned with providing seeds and tools follow close behind essential, life-saving emergency work, with the aim of improving longer-term food security. Such projects often continue, with the long-term objective of re-establishing and enhancing people's livelihoods. For this reason, the implementation of a seeds and tools project as part of rehabilitation is in a crucial intermediary position, linking vital emergency work with sustainable agricultural development work. If this aspect of the function and role of such projects were recognised and taken more seriously, then any potentially negative effect of an emergency intervention on development and development work could be reduced, and even reversed. Suggestions for good practices that support this approach are included in every section of this book. It is every practitioner's responsibility to be aware of this important role and to avoid thinking of a seeds and tools project as an isolated intervention.

Poverty and livelihoods

Poverty, deprivation, and the right to a secure livelihood are major concerns of all aid organisations and their staff. There has been much written on the subject; academic institutions all over the world are conducting research into the root causes of poverty. The definitions of 'poverty' and 'livelihood' vary greatly; there are many different ways of making a living, and in this book we are primarily concerned with rural agricultural livelihoods. Seeds and tools projects are often a first step, after emergency food-aid inputs, towards restoring people's livelihoods. This rebuilding of capacities is a complex process. The changes, sudden or slow, that have made people vulnerable and have put an end to their way of life are, and may continue to be, out of their control. Former social, traditional, economic or agricultural practices may no longer be suitable for sustaining a community. A new set of conditions exist, with different limits, pressures, constraints, and often with some new opportunities. Where there appears to be an opportunity to provide people with the means to a livelihood, that perception may or may not correspond with reality. The complex and dynamic relationship between social, economic, political, and environmental factors must be understood before clear analysis leading to effective action can take place. NGOs are increasingly conscious of the need to involve the beneficiaries in this process of assessment, analysis, and planning if interventions are to be successful.

The conditions that necessitate interventions to distribute seeds and tools vary, and the degree of food insecurity will differ in different situations. To decide that seeds, with or without tools for cultivation, are the most appropriate resources to provide in order to support the rebuilding of livelihoods requires careful assessment and analysis. Assessment can and should be carried out at every stage regardless of the constraints on time, resource or access.

The right to land

Land is an essential resource for agricultural livelihoods. The right of access to land is enshrined in the Universal Declaration of Human Rights (1948). The American Convention on Human Rights and the African Charter on Human and Peoples' Rights also cover the right to property. The convention on the elimination of all Forms of Discrimination against Women (CEDAW) requires states to ensure the same rights for wife and husband in acquiring, owning, enjoying, and disposing of property. The International Labour Organisation convention requires recognition of ownership and possession rights over land traditionally occupied. Property rights, with all their cultural political, economic, and religious implications, vary greatly. What is agreed to in an international forum can be re-interpreted at the national level; there are very few mechanisms for resolving property issues, internationally.

Refugees depend on the good will and beneficent policies of the host country for access to land. It is the mandate of The United Nations High Commissioner for Refugees to work closely with host governments on securing reasonable living conditions, including the right to use land, for refugees. Internally displaced people and returnees rely on their own government for their protection and the recognition of their rights, but government systems of administration may be ill-equipped or politically prejudiced. It is usually the most marginalised groups — the disabled, the elderly, ethnic minority groups, and poor farmers and pastoralists — who are likely to be most severely affected. Women, too, may suffer discrimination with respect to their right to land. NGOs and aid agencies should be aware of current national land policies, existing systems of land tenure, administrative and legal structures, and implementing bodies. (This subject is too vast, and national policies are too varied, to be covered adequately in this book; see the recommended reading list for sources of information.)

Gender roles

Another important issue to be considered is the role of women and men in agricultural production, and related rehabilitation work. Culturally-defined gender roles change very slowly under normal conditions. Communities resist change to well-established social structures and processes, especially when they believe that their very survival may depend on preserving their traditions and way of life. Gender determines responsibilities and activities; access to and control of resources, including land; rights to property; access to education; and influence on decisions in the community.

> *Today there is a growing awareness of women's absolute and relative poverty and inequality all over the world. In spite of the significant efforts of many national governments and at international level, the situation of women has worsened* (Oxfam's Gender Policy).

In sudden or slow-onset crises, social structures may be weakened or disintegrate almost completely, and gender roles may change suddenly and significantly. In fact, in the life of any project, gender roles rarely remain static. In extreme situations, as communities and groups struggle to cope with a crisis, some or most traditional roles may no longer be appropriate. In some cases, family units are fragmented, leaving women, elderly, and very young people as heads of households. People are forced to take on new roles, and be open to new attitudes. A crisis may provide an opportunity to influence change for the better, especially for oppressed groups. At the same time, a crisis, especially a conflict-related emergency, may increase the burden and suffering of women, who may have to take on new roles in the absence of men, but without the social power and rights required for them to perform the new role successfully. NGOs and aid organisations initiating projects in emergencies need to be aware of the implications that their work will have for women, and should take care not to increase women's responsibilities and burdens. Every opportunity should be taken to foster lasting changes that benefit disadvantaged groups so that inequalities are permanently overcome.

Rather than writing one chapter about gender awareness in relation to seeds and tools projects, a 'social relations approach' is taken throughout this book to situation analysis and project development. Reference is made to relevant literature in the list of recommended books. Again, it is not possible to cover this vital topic comprehensively here; it is the responsibility of the individual or team to adopt the 'tools and frameworks' necessary for gender-sensitive project development.

Working in situations of conflict

Increasingly, aid organisations have to work in areas where armed conflict is taking place. Some aid organisations seem to be willing to respond to humanitarian needs in emergencies arising in situations of war, regardless of the dangers. The unpredictable conditions in these unstable environments demand difficult and complex decisions to be made by organisations often lacking sufficient experience and understanding of the context. Along with the crucial issues of physical security of staff, and linked to it, organisations must consider their capacity to operate effectively without compromising their principles and standards of good practice. Most aid organisations set clear limits on the risks their staff should take, and the conditions in which they will operate. These limits vary greatly between organisations; an ill-considered decision or action of one organisation can put pressure on others, especially at field level, endangering the security of all staff. In some countries, efforts to prevent this have been made through close co-ordination and by the establishment of joint agency/NGO protocols and guidelines.

To contribute to the well-being of a vulnerable civilian population without assisting one or both sides in an armed conflict may be difficult, and sometimes impossible. If this is the case, there is a risk that an agency attempting to alleviate suffering and save civilian lives may in fact be providing resources that directly or indirectly are being used to prolong the war.

In the context of seeds and tools projects, there may be a serious problem of the diversion of aid when NGOs and their target populations are the victims of coercion by armed forces that misappropriate humanitarian relief resources. In some cases, such a project may instil a false sense of security that leaves people more vulnerable to violence.

As the dynamics of conflict are better understood it has become clear that, while most people suffer greatly, a few actually gain during wars. These are people, such as the commanders of armed groups, who have the power to control resources, or the transfer of assets from the weak to the politically strong. An economic system may be created in war which promotes this transfer, and is very destructive to livelihoods and community-level subsistence economies. There is then every reason for certain groups to perpetuate conflict as long as it serves their interests. When aid inputs such as food and fuel that are crucial to the war efforts are introduced by aid agencies and NGOs, there is then even more to gain and to fight over. This will increase the likelihood that the intended recipients will be deprived of much-needed resources, which will instead serve to prolong the conflict.

Although warring factions do not usually consider seeds and tools in themselves strategically important resources, these items can become a target when they are the only valuable assets available. Tools can be sold across borders in order to finance the purchase of arms, and seeds can be used as food. Any important resource controlled or manipulated by a group or individual will increase their power. When considering seeds and tools projects in situations of insecurity the benefits must be carefully weighed against any potential negative impacts. (See Chapter 3 for recommendations for project development in areas of insecurity.)

Food security

Some 800 million people are currently suffering some degree of food insecurity, and face malnutrition and hunger. Poverty and lack of education have been identified as causal factors. It has been recognised that there is enough food to feed everyone in the world but some people do not have access to it. This lack of access is apparent at every level. Internationally, there are structures, systems, and attitudes that cause poverty in particular countries; regionally and nationally, there are also structures and systems that create disparity and inequitable access to food; within communities and households, groups or individuals may have differential access to food. In regions where there is conflict and insecurity, it is not only the destruction of infrastructures and disruption of food production that cause poverty and hunger, but the more extreme systems of power and control over resources.

At the community level, there are often traditional systems that provide access to food for the vulnerable, and survival mechanisms that enable people to withstand temporary food shortages. These social and individual strategies may be imperfectly understood by aid organisations, and during times of chaos and crisis, these traditional systems may break down.

Aid organisations have realised the need to understand local food economies in order to deal with problems of food insecurity. But there are no easy answers or quick solutions. Only by assessing the whole situation, and by considering the entire food economy of a community or group, can ways be found that will promote long-term food security. Communities have various sources of food that are utilised at different times of the year in different ways by women, children, and men. The food economy within a community or household can be complex and not easily understood by outsiders, especially during or after a crisis when changes have disrupted this often finely balanced system.

Women are primarily responsible for meeting the daily food requirements for their family, and yet they have often been overlooked as sources of valuable information about the local food economy. In times of crisis women often initiate the more immediate coping mechanisms to obtain food for the family. However, such survival mechanisms are not without cost. They may compromise family health, and the long-term sustainability of livelihoods. The household or community may survive one crisis but in order to do so, may make themselves more vulnerable to the next one.

International economic policies

International and national economic policies profoundly affect local economies and markets. When considering seeds and tools projects, the wider national and international economic context should be considered as part of the initial assessment of the situation.

The World Bank and the International Monetary Fund (IMF) are the international bodies who help to regulate the world's economy. The World Bank's aim is to encourage economic growth and alleviate poverty in poor countries. The IMF oversees the international monetary system, and assists member countries to overcome financial problems. The Bank and IMF may impose structural adjustment packages (SAPs) to improve financial efficiency and enable countries to pay their debts. However, these have often failed to bring about economic growth, and have tended to exacerbate poverty and inequality. The SAPs prescription usually includes cuts in government expenditure, privatisation of industries, the opening up of the economy to foreign trade by removing import taxes and quotas, and the promotion of production for export. Provision of social services, health care, and education is often drastically reduced, and charges introduced.

Small-scale agricultural production may suffer if government credit for agricultural inputs, and technical assistance are cut. Under the pressure to earn foreign exchange, governments are encouraging large-scale extractive projects, such as mining and logging, often at the expense of the environment, and involving the loss of land by agricultural and forest-dwelling communities. In other cases, production of cash-crops such as coffee and cocoa are encouraged, and this can result in a fall in price as world markets are saturated. Countries where large amounts of land are devoted to the production of export crops may be unable to produce sufficient food to feed their population.

In most developing countries, agricultural production is an important source of income to enable governments to repay their debts, pay for imports, and fund general expenditure. The promotion of marketable cash-crops is therefore the focus of many national agricultural policies. While related technological advances have the potential to help poor farmers, it has been the richer farmers that benefit most. This is because of the investment necessary to purchase fertilisers and pesticides needed to grow cash-crops, and the economies of scale of plantation systems. Small-scale farmers may attempt to earn income by growing cash-crops, but this usually means reducing the quantity of food grown for family consumption. Increasingly dependent on the cash-crop for survival, they are very vulnerable to fluctuating prices in the international markets. Cash-crops often require large inputs of pesticide, fertiliser, and water to flourish, and may be destroyed in unfavourable weather conditions. Farmers can quickly become indebted, and eventually lose their land. (See Chapter 9 for further discussion of these issues.)

Many agricultural development projects run by NGOs are now designed to reverse the trend towards cash-crop production by encouraging farmers to grow essential food crops instead. This is not made easy for farmers as much of the seed available in central markets is for cash-crops, and government services such as agricultural credit, grants, and loans promote their production. Farmers trying to maintain or return to more traditional and stable ways of farming may have little support. In planning emergency and rehabilitation projects to assist communities with food production, the implications of this should be recognised.

2 | The project cycle and project management

This chapter deals with the project cycle, and contains two detailed case studies of seeds and tools projects which illustrate the importance of information gathering and careful planning, and how it is possible to take a long-term approach even while responding to an emergency. It is equally possible to prepare for possible emergencies in a situation of relative stability, and the chapter discusses preparedness planning. Finally, the need to be aware of how other organisations are planning to respond to an emergency is discussed.

The project cycle

A seeds and tools project, like any other, will generally follow a logical linear course through its development, implementation, and completion. While it is helpful to think of a project cycle in terms of beginning and end, steps or phases, and separate components, it is never as simple as that in real life. Areas of work overlap, and from necessity, efficiency or common sense, the order of tasks in the project cycle is likely to be modified to respond to the specific situation. For example, in seeds and tools projects it is better to start looking for possible sources of seed and what is available from them as soon as possible, even before the assessment stage is completed, so that information on availability is also part of the analysis. This will help to avoid raising people's expectations only to disappoint them later.

The diagram opposite of a Project Cycle is offered simply as a guide for planning purposes.

The Project Cycle

Project phases

[1] Initial project development or scoping
[2] Assessment, analysis and planning
[3] Project design
[4] Project implementation and monitoring
[5] Project completion and evaluation

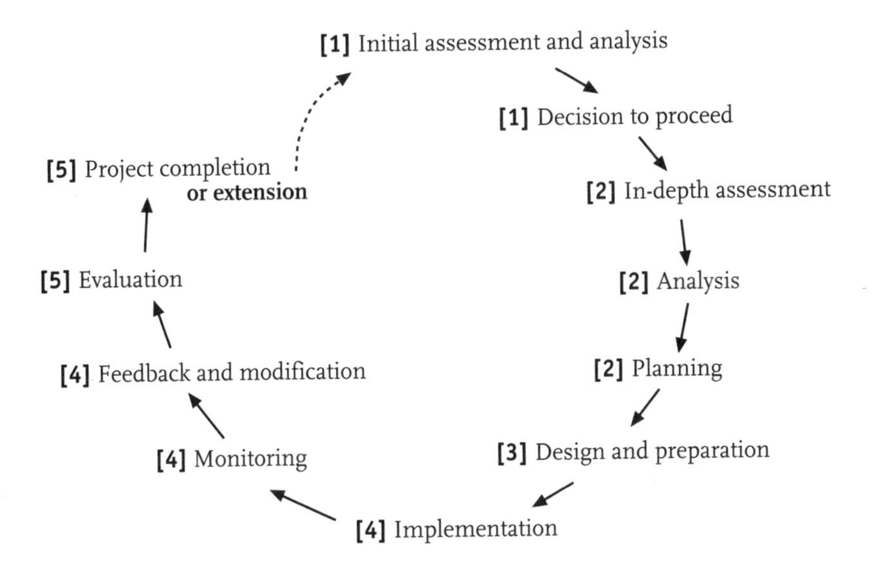

The importance of long-term objectives

A seeds and tools project may be a new project or an addition to an existing one. At the beginning of a new project the priorities are initial information-gathering and analysis, or what is known as 'project scoping'. It is very likely that a year from now, or when this project has finished, your organisation will be at this point on the loop: planning for another cycle with the same project or projects that may have longer-term goals. Even if this is not the initial intention, it is still wise to make short-term plans with the awareness of long-term aims and objectives. This is certainly true of projects involving seeds and tools because of the long-term nature of agriculture rehabilitation and development work.

Learning from mistakes

An Oxfam programme in the Terekeka District of Southern Sudan targeted 7,000 returnee households, who had been affected by the civil war. Oxfam had been supporting various relief and rehabilitation projects in the district since 1985 through a local NGO called ACCOMPLISH. The distribution of seeds and tools took place in May and June 1994. Improved varieties of groundnuts, sorghum, maize, okra, tomato, pumpkin, and cowpeas had been obtained in Khartoum. Axes, sickles, prong hoes, pangas, local-style hoes, and Chillington hoes were also obtained in Khartoum. Germination of the seed was 95 per cent on average.

The evaluation at the end of the project found that food production was low. The reasons given were dry weather, pests and disease, labour shortage at weeding time, the late distribution of seed and tools and inadequate quantities of both, and inappropriate farming practices. The farmers complained that the tools were unsatisfactory, especially the axes and local-style hoes. Other than the resources themselves and a four-month supply of food for the target group, no other benefits were gained.

The appropriateness of targeting only the most vulnerable people was questioned in this case. Recommendations made by the evaluator were:

- Such programmes should have a longer planning period, (one year ahead) to enable information to be gathered and inputs to be procured.
- Food-for-work should be provided ahead of seed distributions, to minimise the likelihood of farmers eating the seed. (In this programme, there were severe food shortages at the time of seed distribution.)
- Seeds and tools programmes should be associated with the provision of pesticides and agricultural advice, possibly using demonstration farms. They should be part of an integrated programme.

A repeat distribution programme was planned by Oxfam in February 1995. The 7,000 households that had benefited in 1994 would receive half-rations of seeds and one hoe. A further 4,000 households would receive full rations of seed, an axe, and a hoe. Learning important lessons from the previous year, field staff made several significant changes:

- This time, delays in seed delivery were minimal, because of longer planning time.

- WFP would be distributing food to most of the beneficiaries at the time of seed delivery.
- Oxfam and ACCOMPLISH field staff were to be trained in extension techniques.
- The Ministry of Agriculture was expected to make pesticides available, and a community-based pest-control programme might be set up.
- Of the tools, 8,500 would be sourced in Juba (in the southern region) where raw materials were available for their construction. The remaining 5,000 pieces would come from Khartoum. Farmers had been shown specimen tools and had approved them.
- Community involvement would be greater because ACCOMPLISH was shifting its focus from Juba to Terekeka, and there was more emphasis in this programme on extension work.

This seeds and tools distribution was also seen as serving a strategic purpose in allowing Oxfam to remain in this area. This programme progressed, not without problems, but the modifications to the project made it of long-term benefit.

This example illustrates some of the problems that can arise when there is insufficient information and planning time. The project was only of limited effectiveness in the first year, but lessons were learnt which were successfully applied in the second year.

Even a very short pre-implementation period can be used to good effect, if the assessment is well-organised, so that crucial important information is gathered and used in planning the project.

Taking a long-term approach to a short-term project does not mean treating it exactly like a long-term development programme. However, the aim should be to strengthen the capacities of the beneficiaries for the future, for example, by the use of participatory techniques in the information collection and analysis stage, and by ensuring that women are well represented. In the design and preparation stage of the short-term project the use of open-ended strategies and objectives can encourage further developments.

Unexpected benefits may be gained when the initial stage of an emergency or rehabilitation project is begun with the intention to support, wherever possible, awareness of and positive actions for human rights, gender awareness, environmental concerns, and capacity building.

A rehabilitation project with long-term goals

In Uganda Oxfam's Kumi District Agricultural Project was carried out in 1991, in response to a long period of insecurity in the area that had disrupted agricultural production and fishing. People there had lost many of their animals to cattle rustlers. Cassava yields had been severely reduced as a result of African cassava mosaic-virus.

The objective of the seeds and tools project was to help people to adapt their agricultural system to the changes that had occurred. It was decided to distribute cassava cuttings, and sesame, rice, sunflower, and vegetable seeds along with tools, four oil-seed presses, and fishing equipment.

Oxfam Uganda was concerned, from the outset, to address the particular needs of women. To this end, Oxfam Uganda:

- Established a distribution system based on sub-parish registers compiled by women leaders, thus using existing local structures and systems that people were familiar with and accepted.
- Developed a flexible registration and distribution system that could be adjusted in response to movements of people, and fluctuating numbers of recipients.
- Distributed items to adult women, this approach having been accepted by the local committees.
- Established good relations with local people, women's leaders, and government representatives in several related ministries.
- Encouraged the collaboration of the local agricultural department in monitoring and extension activities that accompanied the distribution of seeds and tools.
- Developed a project approach and package that would complement distributions made by other organisations and that was considered appropriate by the various stakeholders.
- Recruited women project monitors from the district.

70,010 individuals from the 103 parishes in Kumi district received items from the project. The distributions were timed according to the planting and growing seasons, and most farmers received the seeds on time. Where deliveries were made at the very end of the first planting season, farmers kept the sesame seed for the next planting, on the advice of the agriculturists. Some rice seed was reported to have been delivered too late and was also stored for use in the following year.

This project did not provide everything that the beneficiaries had requested. Farmers had asked for assistance with crop protection in the form of chemical controls. Oxfam decided not to include these in the project, because of the cost and risk of environmental damage.

The project placed a lot of emphasis on the provision of cassava cuttings, to assist farmers to increase their production of this important crop. It was hoped that the distribution of virus-free stems would help to control disease. Along with clean plant materials, the project provided advice to the farmers. The recommendations were: to clean-weed the plot, uproot and destroy infected plants, plant the new stock at a distance from older cassava plants, select only clean cuttings for transplanting, and plant cassava in large blocks. These extension messages were crucial, because the short-term and long-term success of this crop depended on controlling the disease.

Project monitoring showed, among other things, that the method of registration was thought to be equitable and effective by stakeholders; that the relief items and package were considered to have been appropriate by the beneficiaries, government personnel, and committee leaders; and that out of the original estimated 70,349 beneficiaries, 69,604 individuals received seeds and tools. The project evaluation found that the project had been successful in achieving its short-term goal of alleviating hunger and poverty. Farmers reported receiving more advice and increasing their knowledge as a result of contact with Oxfam monitors and agricultural officers. Project assumptions about problems in Kumi had been valid, except for an emphasis on cash crops rather than food crops. Many beneficiaries said that they would have preferred seeds of their main food crops rather than sunflower seeds.

The majority of people thought the Oxfam distribution had made a positive difference to their situation. In the longer-term, the programme assisted people to maintain or expand their agricultural and fishing operations. The cash income was reinvested in agricultural production and the food produced contributed to household consumption and improved nutrition. The longer-term future of cassava in Kumi district was still considered uncertain. Both men and women in discussions expressed strong approval of the strategy to aim distributions at women beneficiaries. Women, because they were given control and responsibility as well as practical help; men, because they had experienced the difficulties of dividing the items from a single distribution between several wives. The local committees supported it because they were concerned to see families benefit rather than individual men, as had occurred in other distributions.

This is a good example of rehabilitation work in an emergency, with short-term goals but implemented with an awareness of long-term outcomes. It was a difficult undertaking that required careful planning, technical knowledge, and a range of skills. Flexibility, close monitoring, and the willingness to modify project plans and methodologies were essential. Such an open-minded approach is important if staff are to benefit from lessons learned during the process.

Initial information requirements

Comprehensive baseline information is not usually available, especially in the initial stage or in emergency situations. There is a list in Chapter 4 of the information ideally required for project or programme planning or scoping, but no situation is ever ideal. The planning cycle is a continuous process in which plans are modified as more information and knowledge is gained. An initial (emergency) assessment needs to provide enough information on which to base a decision as to whether or not to proceed with a seeds and tools project.

To create a project design that has clear objectives, strategies, target groups, activities, resource requirements, and distribution plans requires a great deal of detailed information. While data on agriculture will be the priority, without further information on such things as population numbers, locations, and living conditions it will be difficult to plan effectively. Relationships between agricultural and other activities, and social systems and conditions may not be immediately obvious. In the planning process, these relationships should become clearer, giving rise to more questions. For example, if men are off fishing will they return in time to cultivate? How much of a labour constraint is the human parasite, guinea worm, at planting time? Are the students in school a good target group for the seeds and tools? If information on these matters is available it will facilitate the planning process; if it is not, valuable opportunities can be missed.

Preparedness planning

Many countries fluctuate between stability or near-stability and conflict and crisis. Preparedness planning during a period of stability is never a waste of time. The information gathered during routine implementation of project work is often very valuable. Staff have direct contact with stakeholders, and this period of close interaction provides opportunities for each group to get to know more about the other. Mistakes are made and lessons learned; information is collected formally and informally that will help to predict how the situation might evolve. Very little additional effort will then be necessary to gather the detailed information required to implement a relief project at a time of crisis.

Preparedness plans can be drawn up at any time but fit in well with monitoring and evaluation. They have the same elements as an emergency project plan, with a description of the present situation and the addition of possible scenarios as things might change. Any number of possible scenarios can be included in a preparedness plan if a response to them is possible.

Seeds and tools are not usually a priority requirement in a crisis but are more often included in the rehabilitation phase after people's basic needs have been provided for. Distributing seeds and tools is considered one obvious way of decreasing dependency and increasing food security and self-sufficiency. One important part of preparedness planning is pre-sourcing inputs. In a crisis, there is seldom sufficient time to plan and prepare for project implementation. In the case of seeds and tools, planning and preparation (especially sourcing and purchasing) are often difficult and complex. The more that can be done ahead of time, the better. Having preparedness plans drawn up should speed up another important part of the process: securing funding for the project, especially if plans have been approved by your organisation. (See Appendix 1: Outline of an emergency preparedness plan.)

Co-ordination with other agencies

An important part of the assessment, design, and implementation phases of any project is co-ordination with other organisations. It will be important to know what other NGOs and agencies are doing or planning to do. Time spent working out which organisation will do what and where will be time well-spent, because once projects have been funded it will be far more difficult to modify plans. There may be opportunities to share expensive or hard-to-find resources (trucks, radios, offices, and stores). Information-sharing, especially if your organisation is less experienced in the particular type of work, can be very beneficial. Communication networks and security systems can be strengthened. It will sometimes be appropriate to plan joint distributions with other organisations, of seeds and tools together with essential food aid.

3 | Initial project assessment, or scoping

This chapter deals with some of the fundamental questions to be addressed before deciding whether a seeds and tools project may be an appropriate response to a particular situation. At the very beginning of any project there are four factors that must be carefully considered: timing, appropriate response, capacity to respond, and security. These factors are relevant whether the project will be part of an existing programme; a new, one-off project; or a project that is part of an emergency response to a crisis. There may be very little time to find out and consider the facts before a decision has to be made. In non-emergency conditions there may be pressure to do other work, or pressure because of the short time available before the planting season. Emergency situations have a pace of their own, and usually demand a rapid response. But if this initial assessment phase of a project is done effectively, it should make the decision-making process clear for everyone involved.

It is all too easy to justify rushing into action to respond to desperate needs, especially when there are no other organisations working in the area; and if there are other organisations at work, then it is tempting to think that there must be further work to be done. Scoping is the initial process of quickly assessing the situation to identify the capacities and vulnerabilities of the people affected, and looking for opportunities to help them with their immediate needs while strengthening their ability to meet their long-term needs.

It is a great advantage if the team or individual responsible for conducting an assessment at this crucial stage has experience of the work, close links to the target group, and speaks their language. Chapter 5 covers assessment methods in detail, and gives suggestions for carrying out rapid assessments. Plan the assessment carefully but remain flexible in order to maximise the time spent with the target group. The number of interviews may be small, so strive to make contact with a representative selection of people. These 'key informants' will not only provide information; they will have their own assessment of the situation from their own perspective; their views can be compared and consolidated. A group discussion with a representative cross-section of the population can be very fruitful, but it should not replace individual interviews. Secondary information from reliable sources can also be valuable.

We will now look in detail at four main areas of concern that need to be addressed.

Timing

The timing of an intervention is an important consideration, especially in the case of projects involving seeds and tools, and may often be the main limiting factor. There are many examples of projects that fall short of their intended goals because of insufficient time to plan the project before the inputs are needed, or even when planning time is adequate, because the seeds and tools arrive much too late.

However, seeds and tools arriving late for planting time does not always spell total disaster. A partial, late crop may be better than no crop at all. Make certain that all of the stakeholders know that delays are possible, for they too are investing a lot in the success of the project, and need to be clear that they wish to proceed knowing the risks involved. (For more details on crop timing see Chapter 5.)

Activity calendars

A simple and very helpful exercise at this stage is to draw up a rough sequence of activities in the form of a calendar. This should cover every component of project development up to the actual distribution of the seeds and tools to the individual beneficiaries. The example given below shows that there is just enough time

An example of an activity calendar

ACTIVITIES	Sept.	Oct.	Nov.	Dec.	Jan.	Feb.	March	April
Assessment	XXX							
Assessment report		X						
Project proposal/plan		XX						
Fund raising			XXXX					
Project preparation			XXXX	XXXX				
Purchasing					XXX			
Transport						XXXX		
Storage						X		
Organising distribution						XX		
Secondary transport							XXXX	
Distribution							XX	XX
Planting time								XXXX

for every required project activity before planting time; there could still be a problem with this project if something goes wrong that delays any of these activities. (Allowing for inevitable delays is good practice.) At this initial stage there may not be sufficient information to draw up an accurate calendar but even approximate dates will help to show whether a project is worth considering for this growing season or agricultural year. The calendar can be refined as more information is gathered. The planting time of each crop involved is essential information.

Are seeds and tools the appropriate response?

Before a detailed assessment is done it will be difficult to answer this question. Any baseline information from previous work done in the geographic area or with the community concerned will be very valuable. The many separate factors that must be take into account will be inter-related in a complex manner. Answering the questions listed below should give a clear idea as to whether or not to proceed. (If this is not obvious, then more information about the situation is needed.)

- What population or group are affected? Do you have an approximate population figure?
- Have the changes in their lives left them vulnerable in some way?
- Is or will food insecurity be one problem? Are they receiving or will they receive food aid?
- What means of livelihood did they have before? Have they lost some or all of their usual means of livelihood? How important was agriculture as a means of livelihood?
- What coping mechanisms are they using to survive at present?
- To what extent does the climate and their present environment support agriculture?
- Have the changes in their lives left them without any of the following: shelter/ water/ food/ reasonable security/ hope for the future (this may include rights, freedom and incentives) /basic medical aid/ access to arable land/ the labour (people and/or animal power, fit and healthy) required to farm/ the people with skills required to farm/ personal resources or access to resources including seeds and tools?
- How will they regain any of the things that they have lost? What can they acquire for themselves? How can you help them to obtain these things?
- Will there be any of the things listed above still lacking? If so, how will this affect their ability to farm?

Capacity to implement this project

Your organisation, or partner organisation, on a regional or local level, will probably have a strategic plan or guiding principles for its work. These are intended to focus activities in what the organisation thinks is the most appropriate way. Specialisation may be by sector (health or water) or may target certain groups (disabled or the poorest people). Any potential project should fall within these guidelines. Here is a checklist of questions to help in deciding whether or not your agency is the right one to carry out this project:

- What exactly needs to be done?
- What geographic area and total populations does this involve?
- What other NGOs and aid agencies are working or will work there, doing what and with whom?
- Why is a seeds and tools project necessary?
- Do the following criteria apply?
  - The project is consistent with the organisation's strategy and principles.
  - It is essential in the short term and/or the long term.
  - There are strong links to the community, or there is a local partner who has those links.
  - The skills, knowledge, and experience to implement the project are available.
  - There are adequate resources in terms of staff and access to funding, and the project is a scale that your organisation normally works at.

Here are another four criteria that are more questionable, and would not make a strong case for going ahead with the project:

- It is a gap that needs to be filled.
- There is pressure to do something and this seems like a good project.
- There are funds available which need to be used for something.
- The organisation has decided it would like to get into this area of work.

If this project is on a bigger scale than your organisation is used to, it might be advisable to divide up the work geographically or in some other way, with a partner NGO. Working closely with another NGO on a project or programme of which they have experience and for which they have a good reputation, especially a large project, is an excellent way for an organisation to extend its experience. Try to find out what other organisations have done or are doing.

At this stage, it may not be possible to answer all of these questions. But this exercise will help to identify the information required for a more thorough analysis later on.

Security

In recent years the number of countries that have been affected by conflict originating internally or externally has increased dramatically, and security has become a growing concern for NGOs and agencies. Never before have so many civilians been targets in situations of war. The people who require assistance from NGOs and aid agencies are caught up in the conflict; there are no longer any neutral players. Aid influences politics and war, no matter how hard agencies struggle to remain impartial; and aid workers have become strategic targets for coercion and extortion. In conflict-prone regions the most important questions to ask are:

- Is there a possibility of conflict or insecurity in the area you are proposing to work in? What are the potential risks?
- What is the likelihood of the situation becoming worse or improving?
- What impact could this project have on the local situation?
- Do local people feel secure or insecure? What is the opinion of the local team about the security situation?
- What do other NGOs and agencies think about the security situation?
- Can security systems be established that will ensure staff safety including: communications systems and protocols; clear and comprehensive security guidelines; evacuation plans for all staff; training for staff in communications and security?
- Are there sufficient resources to ensure staff safety including adequate communications equipment and vehicles, including boats and planes for evacuation?
- Is there sufficient information about the security situation at this point to make decisions?

(NOTE: If the answer to the last question is 'no' this may lead to the decision not to proceed because it is impossible to obtain further information or to a decision to gather more security information on which to base a final decision.)

4 | Project information requirements

Having decided to implement a seeds and tools project, the next stage is to gather the detailed information necessary to analyse the situation and design the project. In this chapter, the various kinds of information which will need to be collected during the assessment stage of planning a seeds and tools project, and possible sources of that information, are reviewed, and the chapter includes a comprehensive check-list of information.

The initial assessment or scoping process covered in the previous chapter should have revealed areas where more information is required, or where information is vague and data incomplete. There are no short cuts to information gathering, and it is essential to be well-organised. Taking time to identify what information is needed and how to obtain it is time well spent. This is a stage of the process that may feel frustrating because whatever time it takes to complete it will seem too long. The pressure to begin the implementation of the project will be great.

An example of a seeds and tools project that had good information and used it to plan a successful distribution is given below.

The importance of good information

Oxfam initiated an emergency response project in Sinoe County, Liberia in December 1995, on the basis of a five-day emergency assessment carried out in November by a multi-sectoral team. This assessment showed that agricultural resources were required, along with other inputs. Health information gathered during the assessment and from another NGO indicated moderate levels of malnutrition in children. This supported information gathered from communities reporting food shortages as a result of the very restricted agricultural activities because of the war. People were eating whatever they could forage from the bush. This wild food had been enough to sustain most of them; however, many children and elderly people had died in July and August 1995.

Because of past insecurity, families had moved away from roads and centres, deep into the bush. This made it difficult to assess their conditions and numbers. A survey that included group and 'key informant' interviews with people who had returned to village areas provided data, including a seasonal calendar and a food-economy diagram. The latter showed that approximately 40 per cent of their diet was palm cabbage (the growing tip of the palm tree), which meant that an important renewable resource was being destroyed. The calendar showed that there would be a large food-deficit by January, which would last until July 1996, given the general shortage of wild and other seasonal foods. It also showed that brush-clearing (which had not been done for two years or more), cultivation, and planting times were December/January, February, and March respectively. Seeds and tools were not such priority needs as were medicine and food, but would have to be provided quickly if they were to improve food security in 1996.

Another relief agency, CRS, was planning to provide food aid to the area. The Oxfam assessment results were used in the planning of a limited food distribution from January to June for all farming families in the county. This food was essential if households were going to resume their agricultural activities because they would not be able to gather wild food as they would need all their time and energy to work in the fields. All of the food aid was food-for-work and food-for-farming. Because Oxfam and the other two NGOs involved had no reliable population figures, the seeds and tools distribution was based on the number of farmers, including women, registered to date, plus the number that local agriculturists estimated would register before the growing season. The figure was larger than that used for the health project but smaller than that used for the food distribution. It was almost double that of the former number of registered farmers. However, the estimate proved to be very accurate, because many people who had worked in other sectors such as forestry and small industry were forced to farm full-time again.

Many potential mistakes were avoided in purchasing seeds and tools because time was taken to learn from the past mistakes of other NGOs and to gather technical information from the farmers, local agriculturist, other NGOs, and officials from the Ministry of Agriculture. No evaluation was possible after the distribution because of extreme insecurity in the capital,

Monrovia; but at the time of distribution the only negative feed-back from the farmers was that a few people had not registered in time to get the tools. They were duly registered and given seeds. Monitoring involved spot-checks of registered farmers after the initial tool distribution to see if they had cleared brush and were cultivating land in preparation for planting the seed.

Types of information

There are two broad categories of information: quantitative and qualitative. Quantitative data or information relate to numbers of things: people, acreage of land, yields of crops by weight, etc. This type of information is necessary to determine such things as the quantities of seed, the number of distribution sites, the cost of transport, and the number of field staff required. Some quantitative information may be available from government offices and other agencies but more up-to-date and specific data are usually required. Reliable data can be gathered using various statistical research methods, such as sampling a small percentage of the total population throughout the geographic area. Data can also be obtained from or calculated from vaccination records, registrations, administrative records, health records, and similar sources.

Qualitative data do not involve exact measurements. They relate to things like traditional practices, gender roles, existing social structures, the acceptance of ideas and plans, and can also be collected using surveys. Both types of information are important and are needed for analysis and project-planning.

Sources of information

There is usually a wide range of possible sources for information, each with its specific advantages and disadvantages.

In stable situations there will probably be government institutions that can provide maps and statistical data, which may or may not be up-to-date, on agriculture, populations, and local markets. Quite often, such information is regionally oriented and not specific enough for local projects, but is useful general background information, which can be cross-checked with information from other sources.

UN Agencies and ICRC, when present, can be very good sources of information. Their data may be more up-to-date and less centralised, depending on their previous and present work. Other aid agencies and NGOs that have been working in an area for some time can usually provide information. (The quality, of course, will depend on the organisation's standards and experience, so it is wise not to rely on it too heavily.) Your own organisation's baseline information will be invaluable even when it is from a period before the present crisis. Local NGOs and grassroots groups can obviously be very valuable sources of information and expert knowledge; particularly in terms of qualitative information.

Aerial observation and satellite images are two very 'high tech' means of observation, which have some advantages and may be available for particular areas. Both require first-hand knowledge of conditions on the ground and can then provide information about the situation in a wider area.

Maps of the geographic area and greater area will be very valuable. They may also be useful during interviews. Local authorities and co-ordinators may have maps or may be willing to draw maps for you that show features such as schools, health centres or farm fields. (Include drawing materials in the list of assessment supplies required.) Some communities use 'hours walked' to indicate the distance between two locations rather than kilometres or miles. Distances may prove to be important when planning distribution sites as people will need to travel, walking in most cases, from their homes to the site and then back, carrying whatever resources they receive. Check the sites of clinics, schools, markets, and community centres; these may prove to be the best distribution sites because of their central location.

Gathering data from as many different sources as reasonably possible and cross-checking or comparing them is advisable in any situation. Remember that one of the best ways of gathering information is the most informal one: simply talking to people. In both formal and informal meetings, information can flow both ways. Never forget that the beneficiaries also require information: the success of any project will depend on this. Every opportunity should be taken to share relevant information with them.

Check-list of information needed for assessments

Initial and emergency assessments should try to cover the parts of this checklist printed in bold while baseline surveys should try to cover it all.

Structures:

- **Form of administration for each group or community area; local Administrators and Sectoral Co-ordinators of Agriculture, Health, Water, Education.**
- **Local leaders and traditional structures and how they relate to the official Administration.**
- **Committees, NGOs and agencies working in the area; their activities, and the possibilities for co-ordination and co-operation.**

Population:

- **Population figures with detailed breakdown of residence, displaced people, and refugees; ethnic and geographic origins; livelihood groups**

(fisher people, agriculturists or pastoralists); migration patterns, past and present; number of children under five, and by age-group and sex.

(Note: Population information is the most difficult to gather; it is both time-consuming and contentious. Rather than wasting time because you do not agree with official or other sources it is better to find a method or formula for rounding up or down the figures you are given and if possible checking them with other sources, like health programmes.)

General situation:

- **Security information for each geographic area, group, or community (from several sources, if possible). The likelihood of any change in the present situation.**
- **Relations between different ethnic groups, residents, and displaced people.**

Shelter:

- **Shelter: Are there any problems with shelter; if so, for who and why? What resources are available for constructing shelters.**

Water:

- **Water quality** and quantity, **sources, accessibility for each group.**

Health:

- **Mortality, and the five main causes of mortality.**
- Morbidity and the five main causes, and vaccinations done.
- **Epidemics (source: community health professionals).**
- **Local health structure and services**; local support.
- Use of traditional healers and medicine.
- Nutrition, malnutrition rates (where time permits this could be co-ordinated with a nutritional-status survey).

Education:

- **Education, numbers of schools, teachers and pupils in each grade;** school enrolment, by sex, calendar-year; educational materials available, community involvement.

Logistics:

- **Types of transport; roads, bridges, airstrips (with seasonal changes); communications; storage facilities; distribution points; border crossings and checkpoints.**

Livelihoods and the food economy:

- Types of livelihood
- Food sources and types, the main food economy including:

  • Agriculture: crops grown and crops which could be grown; seasonal cropping calendar; quantities usually harvested per house-hold per harvest per year; acreage usually planted and to be planted by household head this season; crop varieties plus quantities of seed present and used per crop per acre; constraints (seed shortages, pests, diseases, soils, labour and climate); types of tools and equipment used, usual sources and present supply (get samples); system and availability of farm labour; traditional farming systems; storage facilities; farmer's groups or co-operatives.

  • **Animals: animals raised and numbers of each per household head;** number killed for meat/sale of meat; number sold/traded per year; number that died and why; who eats the meat. **Milk production per species for each season;** who has access to milk, **what quantity. Constraints to animal husbandry** (pests, diseases, grazing, fodder).

  • **Fishing: Fishing season/s, who fishes, where** and how the fish is processed; who eats the fish; **constraints to fishing past and present.**

  • **Wild Foods: wild foods eaten, past and present**; where they are gathered, when, by whom; **quantities eaten daily** by whom; preparation by whom; quality of wild foods as perceived by household head; **constraints and limitations.**

  • **Other sources of food: foods available, from where; quantity eaten daily, by whom; relief foods received, how much and by whom; quantities and types of food available through trade/purchase, eaten by whom.**

- Sources of seeds

  It is important to establish whether or not there are local sources of seed accessible to this population or to the project, and the varieties and quantities available. There may or may not be a choice between locally produced seed and commercial suppliers. (See Chapter 7 for details on seed types.) They each have their own benefits and drawbacks. If one source can not provide the correct varieties or substitutes that are appropriate for the growing area, or in the quantities required, it may be necessary to use several sources. It is important to know this as soon as possible because sourcing seed can be complicated and time-consuming.

- **Tools used**

  Tools are quite often adapted to local conditions and user specifications. It is important to find out what these are as soon as possible. Many projects assume everyone uses the same tools from area to area, but experience shows otherwise. Ask for samples of the tools required by each group. (See Chapter 7 for more details on tools.)

5 | Assessments

Having decided on the information required, it must be collected and analysed, and this assessment process is the subject of this chapter. There are a variety of approaches to information gathering. There is nowadays less emphasis on the collection of quantitative data using statistically viable methods, an approach which can be very time-consuming and expensive. Instead, qualitative information is recognised as being equally important. Some standard methods used in conducting assessments will be outlined in this chapter. Many NGOs and agencies have, however, developed their own methodologies that focus on certain criteria based on a particular set of values and assumptions.

A food-economy approach

A holistic approach to food-related assessments will provide a more complete understanding of how people cope with changes that cause food insecurity. All the possible sources of food before and during a crisis are investigated, and also the vulnerability and means of access to food of each group within a community. Crop production is usually a major factor in food security, and considered in this wider context of the 'food economy', the question of whether seeds and tools are the most appropriate input at the time can be more clearly addressed.

Taking a food-economy approach to assessment and analysis means that not only is information on every source of food gathered, but also information on past trends. (For example, how many bags of sorghum does a good farmer harvest in the best year, average year and in the worse year? Was last year a good year, average or bad year?) This information will not be relevant in the context of a refugee camp but could be important if displaced people can develop livelihoods similar to those of the local people. In that case information will be required from the local community, to establish whether or not there are chronic or occasional food shortages. Information should also be obtained on other sources of food, such as wild foods, fish, and purchased grain which make up for shortages in milk production, crops, or meat. This reveals the opportunities or capacities and the weaknesses or vulnerabilities of the food economy in a particular environment at a particular time.

Training may be necessary in the assessment methodology used for this food economy approach. SCF and WFP have promoted this approach and have provided training to NGO field staff in some countries.

Participatory methodologies

Methodologies which provide a framework for gathering and analysing information in a participatory manner also promote a deeper, holistic approach because they place great emphasis on involving the beneficiaries: the people who know the situation best. A 'social relations' approach focuses on analysing existing inequalities in the distribution of resources based on aspects of social identity such as gender and ethnicity. Techniques have been developed to involve the interviewee as much as possible in the assessment and analysis, facilitate clear communication and understanding between the two parties, retrieve information that is easier to cross-check, and make the interviewing process much more interesting for everybody.

Participatory Rural Appraisal or PRA is one well-known assessment 'package' which uses some old and some new techniques. Its participatory approach and methodology work best in situations where the assessment team has a relatively close relationship with the community or group. Rapid Rural Appraisal is another package that has been developed to avoid bias in the research work and support decision-making when time is short, as in the case of emergency assessments. These methodologies have their own problems and continue to evolve. They include techniques such as:

- Semi-structured interviews
- Group interviews
- Sketch-maps, transects, diagrams, calendars, and other visual aids
- Ranking exercises
- Role-playing exercises
- Informal workshops
- Direct field observation

These techniques, and others, are described in a wide range of publications on each methodology; see the list of further reading at the end of the book.

Assessment objectives

Emergency assessments, conducted in situations of crisis and conflict, must be done quickly and efficiently; good organisation is essential. Deciding on the main objectives of the assessment is the first step. Possible objectives might be:

- To identify the vulnerabilities and capacities of the target group focusing on human resources, natural resources, and social structures.
- To identify and prioritise the immediate needs, concentrating on water, shelter, healthcare, and food.
- To identify logistical feasibility and needs, plus security levels.

It is very difficult to determine levels of malnutrition or find evidence of food insecurity in a short time. Any evidence about these indicators should be carefully compared to information on the capacities and present resources of the group or community and the constraints and limitations they face. There may be a link between food insecurity and the loss or lack of resources such as seeds and tools; or food insecurity may result from general insecurity due to conflict. When distribution of seeds and tools is part of the solution to food insecurity you will need to know exactly when they are needed for cultivation and planting, and approximately when the crops can be expected. Logistical considerations are crucial; the feasibility of the distribution depends on the availability of transport, good roads or airstrips that remain open for the period up to the growing season, adequate bridges on primary and secondary roads, and storage facilities which are large enough, safe, and very dry.

In baseline assessments there will be more time to achieve your objectives in a participatory way. It is always important to include the community not only in the provision of information but also in identifying and prioritising problems, constraints, and limitations as well as analysing the situation. The objectives of a baseline assessment might be:

- To obtain background information on economic, social and geographic conditions.
- To identify the main vulnerabilities and capacities of the community or group looking at human resources and skills, social infrastructure and natural resources.
- To identify and prioritise, with the people involved, their most important needs.

Preparing for an assessment

Government officials, local leaders, groups, and individuals who will be involved in the assessment directly or indirectly should be informed in advance when the assessment will begin, the intended schedule, and the aim or purpose. In some situations it may be necessary to obtain approval or official permission to carry out an assessment; but in any case, informing people of your intentions in advance is a matter of courtesy and always good practice. Official letters from your office or government authorities may be required, and are always helpful and a sign of respect to local authorities and leaders.

The assessment team

Ideally the individual or team conducting the assessment will have had previous assessment experience. When time is short, as in an emergency assessment, there will be little opportunity for training or practice. If there is a lack of experienced personnel, then at least the team leader should have experience. Additional team members could be counterparts, and technical staff or project managers with field experience. A team consisting of members with knowledge of various sectors is always helpful. Gender balance is important and will determine the quality of information gathered. In some situations it may be necessary or advantageous to have a team consisting of people from two or more NGOs or agencies, to provide better coverage of sectors such as health, nutrition, and water. The team may require at least one translator or guide and local agriculturist.

The following qualities are desirable in team members: patience and perseverance; the ability to listen; a sincere interest in people; a calm and composed manner; the ability to communicate clearly; good powers of observation; the ability to relate to a range of different people and to be relaxed and natural; a care for details and accuracy; the ability to negotiate and be tactful; natural curiosity and interest in the environment; the ability to be assertive when necessary but not aggressive, and to be innovative, imaginative and spontaneous. People do not respond well to those with an air of superiority or self-importance. A mixture of confidence and genuine humility will often break through any barriers quickly. Remember while you are busy observing people they will be watching your team and its behaviour. Just as no one person in a team will have knowledge of all sectors of work, nor will any one person possess all of the desirable qualities listed above; a well-balanced team makes up for this when it works closely together.

It may be possible to provide a range of training for staff, such as a RRA or PRA training course, a workshop in assessment methodology and techniques, or simply the opportunity to practise interviewing people. Some agencies and NGOs have developed their own assessment methods and training courses for staff, and may make these available to other NGO staff upon request. Training staff who will be involved in future assessments is worthwhile; even experienced staff can pick up new approaches, techniques, and ideas that will increase the effectiveness of their work. Ideally a combination of formal training and on-the-job training facilitated by well-experienced staff should be offered.

A short team-meeting should be held as soon as possible, to decide who should act as team leader; who is responsible for the various sectors, including logistics and security; the geographic coverage and schedule of the assessment; what is needed and who will provide it; what co-ordination is required with other NGOs and agencies, government, and local administrators; other support required and responsibility for providing it (for example, security, communications, food, accommodations and transport).

At this meeting, the team should decide on methodologies and techniques, the formulation of questions, the format for the recording sheet, how to review existing information, and whether training or practice in interviewing techniques is required.

Survey forms with lists of questions and blank spaces for the answers are sometimes used to standardise information. It may not always be appropriate to use these forms to dictate the structure of every interview; this can be very tedious for everyone involved and therefore counter-productive. A simple checklist of the information needed can be used instead, to focus an unstructured interview. However informal the interview, the use of standardised recording sheets, one for each interview, is recommended. These categorise information, make it easy to see what is missing, and make report writing much easier.

Carrying out an assessment

Preliminaries

On arrival in the area, introduce the assessment team and explain its objectives to counterparts, community leaders, local support personnel, key informants, and other interested individuals. Take time to discuss:

- the general situation, and any security problems
- the objectives of the assessment, the need for first-hand information, and the use of the information
- methodologies to be used, the intended schedule and logistical needs, such as women and men to act as translators or guides
- the fact that this is only an assessment with no guarantee of any response
- the importance of communication with all the various groups within the community.

Local assistance

The selection of local translators, guides and support persons can be done at any stage, preferably by counterparts, or when group discussions are initiated with traditional leaders. This will allow selection to be based on observed language and other skills. Make it clear how much, if anything, local personnel will be paid for their services. Discuss with them:

- the assessment objectives and use of gathered information
- the interview approach and techniques to be used
- the checklist and questions to be used in the structured or semi-structured interviews ensuring that the translators understand the terms and concepts
- the importance of ensuring direct translation
- the importance of not encouraging expectations.

Information sources

There are several groups of people who may be useful sources of information:

- counterparts, authorities, sectoral co-ordinators
- traditional community leaders, elders, including women leaders
- key informants for each sector i.e. fishing, farming, hunting, trading, etc.
- teachers
- religious leaders
- health-care workers
- A selection of community members representing different groups including: vulnerable groups and or displaced people, age, gender and ethnically defined groups, women through semi-random interviews in their homes. It is vital to interview women in the community, because of their role in food production, gathering, storage, and preparation, and water and fuel collection.

In an emergency assessment there may only be time to interview authorities, local leaders, and key informants.

Interviewing

Interviewing techniques can make or break the assessment. Imagine how long you would be happy to sit and answer a stranger's questions that seemed silly or obvious to you, especially in a crisis when you don't know where your next meal is coming from. It is up to the assessment team to make interviews as interesting and as pleasant as possible. The reward will be reliable data.

The team will be interviewing groups and individuals using methods that have been agreed upon as appropriate for the time-frame and situation. Two valuable techniques in agricultural assessments are mapping and seasonal calendars, and these can be introduced as part of group and individual interviews.

Interviews should not last for more than 30 minutes; politely cut an interview short if you are not gathering useful information. Be sensitive to the fact that you are entering a family's household space, and always introduce yourself. Let the group or individual talk about what interests them for at least some of the time, as this may reveal some important information, which can be followed up in future interviews. In group interviews note differences of opinion and consensus. Information about trade or exchange with other groups and communities, even at what seems like a long distance, is important as this may indicate sources of food, seed, tools and other vital resources.

It is usual to interview in pairs, one person recording while the other leads the interview. Women are most appropriate, and in some situations, essential, for interviewing women, both in groups and individually. Men should be excluded from interviews with women if possible.

Do not make any promises or encourage any expectations, and avoid political discussions and questions about military activities unless they relate directly to security matters which might affect the project.

Field observations

While conducting an assessment the team will constantly be observing people and their physical surroundings; people's reactions to their surroundings, to other people, and to the present conditions and the presence and questions of the assessment team. The information collected in this way will be qualitative (unless you actually measure, count or weigh things in a systematic way). In quantifying things like grain in household stores, any information from sampling is only statistically representative when a statistically valid survey is set up (or every household is sampled). However, qualitative information may be adequate for assessment purposes; for example, if all of the households visited had no grain left in store, or if on most farms observed cultivation had already begun.

A checklist of useful observations will help to structure the process:

- water sources, type and number; were they protected, were there queues, what is being used to carry water, and to store water in the household;
- the general appearance of the population, and of specific groups, such as age-groups
- the conditions of shelters
- the existing resources in the environment for creating shelter
- the condition of the people's clothing
- the condition of and presence or lack of blankets, mosquito nets etc.
- food and food sources; crops, fish, wild foods, foods in the market place, food aid in stores and households
- daily food preparation and consumption including brewing
- assets; agricultural tools and seed, fishing equipment, cooking pots and utensils, cottage industry tools and equipment
- food-storage systems and facilities at every level
- livestock; types, numbers at household head level
- local markets, activity levels and who is buying and selling
- logistical details; road and airstrip conditions, condition of storage facilities, and conditions at potential distribution site.

Team members will observe different things, and the same things in different ways. There may be time to make notes of some observations but not all, and there is always the risk that these will be forgotten and lost. To avoid this, the team should meet at the end of every day to compare and record their observations. Planning time for this important daily activity in the schedule will avoid its becoming a burden and not being given the attention it deserves. Team members will no doubt be tired after a day of field work; if the recording of observations is combined with a tea break and a chance for each team member to share how they feel and think about the day's activities, it could then become a pleasant round-up activity to end the working day. A list can be compiled of things to check on the following day; for example, an activity or practice whose purpose is not understood.

Detailed information involving measurement may or may not be required. It may be sufficient to check measurements of a small number of fields against what farmers are reporting. Experienced agriculturists may be able to estimate acreages from visual observation. When accurate measurements are needed, remember that most farm fields are not square. This will make calculating areas more difficult. For this the team will need to draw diagrams of each field with all of the measurements taken.

Areas of land may be measured in hectares, or acres; it is important to know the unit of measurement used by the community. Long tape measures can be used for this work but it is easier to calibrate every team members' stride. This is simple and can be done before you go to the field.

Simple systems of measurement

Each team member can walk as they always walk between two markers set at 60 ft. or 20 metres apart. Count the paces to the 1/4 pace and divide them into 60 ft. or 20 metres. e.g. 24.5 paces into 20 metres. = 0.816 metres. This will give each team member their own factor to use to measure distances when pacing out the perimeter of a field or even a store. (See Appendix 2 for area measurement conversions.)

Local measurements for grain, seeds, salt and flour may be made using local standard common containers. These might be tins in which some popular commodity is sold, such as vegetable oil or tomato paste. It is simple to convert those units into international ones, e.g. one oil tin (large size) = 7.5 Kg. or 16.5 lb. Asking a local person (an agriculturist, teacher or health worker) about local systems of measurement before you start the field work can save valuable time. The team may need a weigh scale for field work, especially if little is known about the system of measurement.

Random sampling

Household-level surveys and sampling are ways of structuring some of the interviews in an assessment so that information gathered from a small percentage of a population will represent the whole. Households in villages and camps can be randomly chosen using a simple technique. Walk to the centre of a group of dwellings and spin a bottle to find a direction for the first sample, then again at that location to point to another house. Continue until you have reached a pre-agreed number of households or time span. Try to avoid letting local people select households as they may have some bias that they are unaware of. If a nutritional survey, using statistically viable procedures for household sampling, is taking place at the same time, then it might be possible for those doing the survey also to collect any statistically representative data which you require.

Remember information collected from an emergency or baseline assessment only provides a small snapshot, not the whole picture. Be aware that there are things you still do not know and take every opportunity, during your work, to increase your knowledge of the situation.

Making maps and calendars

These activities can be done as part of an interview with a group or individual or as completely separate activities.

A map can be drawn with a stick in the sand or soil, with rocks, twigs, leaves, and seeds used to represent landmarks: villages, schools, rivers, bridges. This map will be visible to everyone present; it can be used to stimulate discussion and can be easily modified and extended as the group discusses it. One team member can make a permanent record of it at the end of the interview.

People will have differing perspectives and specific information to contribute to such a map. After making a general map with a cross-section of people, to get as much information as possible, appropriate people can be asked to provide specific information about, for example, fishing spots on the river, locations of working and broken water pumps, or the location of particular crop fields. Discussions arising from the exercise may reveal important detailed information.

Seasonal calendars which focus on farming or food security activities are invaluable for identifying periods of hunger or food insecurity. They can be drawn up by groups or individuals, by a cross-section of the community or by a specific group. The process of creating them with people is interesting and stimulating. The calendar can be drawn on the ground, using seeds and other items to represent different crops and different activities. Use the seasons suggested by the group, and start where they naturally want to start. Encourage discussion and make modifications to the calendar in the light of what is said. Women and men will have separate activities and responsibilities which should all be included in the calendar. Activities that may not be directly seen as farming should be included, for example, burning brush, fence making and tool making. Activities that are not farming but require labour and may compete with agricultural activities can also be included, for example, honey collection, food gathering, pot making, roof making or repairing, fishing, basket making and attending school. Specific information on the times that each variety of a crop is planted and harvested is essential. Such a calendar may become complicated and full of detail, so care and patience will be required to get it right. Often farmers are more than willing to take the time to do it properly because farming and other activities are very important to them. Show some of them the copy on paper of what was drawn on the ground to check the accuracy.

Exit or final meeting

It is important to talk to and thank your hosts at the end of the assessment in each area. You may not see them again for some time, so explain clearly what your intentions are, the process that will take place after the assessment to decide what you will do, if anything, and when they can expect to hear from you or see you again. It is important to give truthful answers rather than make vague promises that your organisation may not be able to keep. It is possible to show personal concern while maintaining professional objectivity. Adopting this approach helps to gain the respect of the community.

Information analysis

Under ideal conditions most of the assessment and analysis would be done as a consultative, participatory activity. The vulnerable population, represented by a cross-section of individuals and groups, would provide information about the main problems, constraints, and priorities, their analysis of the situation and the appropriate response. In close consultation with the NGO or agency, they would decide what realistically can be done, given the available resources. The community would supply skills, labour and resources, with the NGO filling any gaps to facilitate the project and ensure success.

This ideal scenario is rarely achievable. Time constraints may partially or completely eliminate any opportunity for participatory group analysis, but individual analysis may be obtained from a small number of respected knowledgeable people qualified to represent a certain group within the population, for example, the leader of a women's project group, or a leading farmer or fisherman.

Organising information

Having gathered as much information as possible, given the time and resources available, the team will need to analyse it, make some planning decisions, and then design the actual project. Analysing information and making appropriate decisions does not require special expertise; but a lot of data may have been collected and it is therefore necessary to organise them before it is possible to begin to analyse them. If information checklists and standard recording sheets have been used, and maps and seasonal calendars created, then the data gathered are already categorised to some extent. (An example of an assessment report format is given in Appendix 4.)

Cross-checking information

The first step is to cross-check information from the various sources. The assessment team needs to compare data and decide what they will accept and use. Ideally sets of information obtained from different sources will validate each other.

An example of comparing between sets is to compare the figures for acreages planted by farm families as given by the families, local authorities, agriculturists, and from your observations. Examples of comparing between different sets in a category are: comparing quantities of seed requested and acceptable seeding rates with acreages proposed by households or groups; or the estimated numbers of farm labourers and the average rate of cultivation, with proposed acreages of crops. An example of comparing between sectors or categories is comparing the total number of children under five years obtained from vaccination records with the number obtained by calculation, using the percentage of this group within an average population and the total number of farm families.

Comparing information between geographic areas or different communities or groups may reveal some interesting differences or similarities, and indicate levels of vulnerability. This may help to decide priorities or how to distribute limited resources according to need.

If sets of data are very different, the cause could be poor communication between the assessment team and the interviewees; different perspectives, motivations or agendas among the various sources; or a very chaotic situation where people are really just guessing at the answers. When in doubt remember the closer you get to the individuals or families concerned the more accurate the information should be, especially if it is successfully cross-checked by observation. When information from one source or even one area assessed is of questionable reliability, and there is no time to go back and check, it will have to be ignored, and decisions based on what are considered accurate and reliable data.

Misinformation

False information is a difficult issue; there are several reasons why people might deliberately give information that is not correct. Where food aid is involved or could be provided along with other non-food inputs, population figures are crucial. Potential recipients may assume that they will only receive a portion of the food they ask for. If they inflate the number of people at risk, then they may be provided with sufficient food for everyone. The fact that some of the food aid is likely to be diverted by a group in authority may also lead interviewees to inflate population figures. Even when there is no possibility of food aid being provided people may hope there will be and provide exaggerated figures for farming families. In situations of extreme insecurity or conflict (past or present) one can imagine the pressures on people to get what they can while they can.

Seeds and tools are not usually contentious as a resource but the team may become caught up in the politics of food aid. If it is clear that incorrect data has been provided and it is related directly or indirectly with present or potential insecurity among the target group then the team should seriously reconsider providing seeds and tools to this group at this time. Agricultural projects and programmes require a certain amount of long-term stability in the area for success.

Analysis framework

It should now be possible to answer the 'big four' questions listed in Chapter 3 concerning timing, appropriate response, capacity, and security, and to decide on objectives. An analytical framework may be helpful to guide this process. Each framework usually supports a specific approach that reflects a set of values and assumptions and will provide a set of procedures, techniques or tools. These focus on analysing factors that directly relate to that approach. For example, Oxfam produces a guide to frameworks for gender analysis which offers information on them that will help you to choose which one to use. The framework selected will to some extent influence the project design and interventions, therefore the organisation's policies, aims and objectives must be carefully considered when choosing a framework. Some standard frameworks that are widely used (for example, logical framework) focus on efficiency and effectiveness of resource allocation. However, it is possible to work effectively without using a formal framework.

A decision not to proceed

There is always the possibility that, after analysing the available information, the team will conclude that it is not appropriate for the organisation to initiate a project. If so, it is essential to inform the people that will be affected. This will allow them to get on with their decision making, planning, and organising. It may help other organisations to make difficult decisions. A 'no' decision will be respected if it is clear that it is based on sound and thorough reasoning and analysis.

The assessment report

Whether or not a project is recommended, an assessment report will probably be required. The report will list the main recommendations, along with the information collected, the analysis of the situation, and a probable evolution. It is an excellent supporting document for any project proposal. (A format for an assessment report is given in Appendix 3.) The format of the report is similar to the information checklist but includes a section on background or what led to doing the assessment; type of assessment (emergency, baseline or other), and methodology, including techniques used and any limitations encountered; the composition of the assessment team; the schedules and specific objectives of the assessment; the situation analysis and probable evolution of the situation; assessment checklists; and the recommendations. Samples of any forms, maps and calendars can be included as appendices.

6 | Project design

Having received confirmation that your organisation or donor is interested in principle in the project, it will be necessary to submit a detailed project proposal. This chapter describes the process of drawing up a proposal to be submitted for funding approval. This proposal will in effect be a project plan: what is to be done, why it is necessary, who will benefit, what separate activities will be involved, when and how these will take place, and also how the success of the intervention will be measured.

The information required for the process of designing and writing a plan for a seeds and tools project is:

- All the available secondary information and relevant material, the assessment information and report.
- Your organisation's policy statements and strategic plans.
- Maps, and aerial photos and satellite images of the area if available, and other logistic information.
- Technical information on seeds and tools.
- Lists of possible sources for seeds and tools.
- Copies of past seeds and tools proposals, reports and evaluations.
- Minutes of all meetings directly relating to this project especially meetings where decisions were made.

You may also need background and reference material on assessment and analysis frameworks, on gender awareness, capacity building, environmental awareness, and human rights issues.

Writing the project proposal and plan

Most donors have a format or checklist of information they require in a proposal, and most aid organisations have their own format for plans and proposals, which ensure that all relevant information is included, and facilitate internal review. A generic format is provided below. However, during initial thinking about the design of the project, the team's thoughts and ideas may naturally follow a different order, and this creative flow may be restricted if a formal structure is imposed. The final format can be used as guide while the team develops plans in a way that is more comfortable to them.

Any planning process needs a starting point; this can be the main problem as identified in the assessment. From this a plan can be developed by writing down a framework of supporting ideas and then filling in the details.

There are various methods to facilitate the design process, for example, a Problem Tree or Mind Mapping. These methods allow ideas to flow and unfold so they naturally create their own structure. An individual, a small team or a group of people can use these methods, which are very visual with boxes, circles and lines linking ideas and directional arrows drawn to indicate relationships. (See Appendix 4 for an example and explanation of a Mind Map.)

The team leader, project manager, or a group of staff may take on the responsibility of writing up the proposal, using the informal guidelines resulting from the collective design process, with the main objectives, activities, and strategies clearly defined.

Contents of a project proposal

Introduction and background

The introduction may include the circumstances that led up to the proposed intervention: a brief history from your organisation's perspective; what involvement your organisation may have had in the past; what alerted you to the problem; how the problem was investigated; any initial assessments, methodology used; present position, and proposed actions.

The background information usually consists of an account of the actual situation, showing what led to the present crisis. This can include information about geographic areas, ethnic groups, population figures, political events, climatic and economic conditions and changes.

Aims and objectives

The aim will be a clear statement central to the problems involved, for example 'to create food self-sufficiency', 'to improve food security' or 'to secure or improve livelihoods for a certain group of people'. The main objectives will be statements of intended achievements that will support the aim. Objectives are not activities, like 'providing seeds', but are goals such as 're-establishing former levels of agricultural crop production' or 'enabling families to establish a farming livelihood'. There may be subsidiary, specific objectives necessary to achieve the main objectives, for example 'ensuring that women and men receive information on planting and growing vegetables'.

In addition, you may wish to show how these objectives relate to your organisation's policies or strategic plan. Certain objectives may directly or indirectly address important issues such as women's rights or environmental degradation. You should explain clearly what it is that you want to achieve and exactly how your project will achieve it. Be realistic in setting objectives; there is no point in taking on more than resources, skills, and the present situation will allow for.

Indicators

Indicators are linked to the objectives and activities. They will be used in monitoring the project's progress and in any evaluation. Appropriate indicators can be broadly identified while the team is developing the aims and objectives, and developed in more detail when funding is assured, when the objectives are finalised, and when the beneficiaries can be involved and the details of project implementation are clear.

Indicators should be selected carefully to represent the various interests of a cross-section of stakeholders. They should be measurable, reveal significant changes that occur, and match the project focus and strategies. Indicators can be quantitative (numbers of acres of crops grown) or qualitative (reopening of food markets in the project area); they can be direct (bags of grain produced per family) or indirect (prices of grain drop in the local market).

An example of aims, objectives and indicators in a seeds and tools project

Aim Reduce suffering from hunger

Objective Improve food security

Specific Objective Provide all families with seeds and tools

Indicators

1. Levels of malnutrition reduced.
2. Dependence on food aid reduced or eliminated.
3. Number of families growing crops.
4. Bags of grain harvested.

Notes:

1. Indicator 1 is linked to the aim, it may involve another intervention (food aid) and it may be expensive to monitor, by for example, doing a nutrition survey. Indicator 2 is linked to the objective, it is indirect and qualitative. Indicator 3 is linked to the specific objective, it is direct and quantitative. Indicator 4 is linked to the objectives, is direct and quantitative.
2. This is an example of a 'logical framework'. Setting out aims and objectives in this form may be a requirement of funders.

Strategies have been defined as pathways that will be pursued to move towards the realisation of aims and objectives. An example of a strategy to achieve increased food production is 'promote extension work with women and men farmers in effective methods of crop management'.

Project activities

The activities necessary to establish the project and reach the objectives will be everything from sourcing, purchasing and transporting seeds and tools, and the distribution of those seeds and tools, to writing a final report. A schedule of these activities will help to determine short-term and long-term staff requirements, transport or monthly vehicle needs, and the timing of key activities. Extension activities should be included here, for example, 'provide informal training and information to the beneficiaries' when the project is supplying seeds or tools that are unfamiliar.

Target groups

Defining who exactly should benefit from the intervention will involve decisions about who is to be targeted, why this or these groups have been chosen, the numbers estimated in each group, how they will benefit and how others may benefit. An explanation as to why assisting these people fits in with your organisation's strategic plan or policies should be included in the proposal.

The targeting of specific people within a group or specific groups within an area or region may be necessary for a variety of reasons. If there are limited project resources and everyone is equally vulnerable, then you will need to decide whom to target in order to bring the greatest benefit to the community as a whole. Two approaches can be adopted; they are not mutually exclusive. The first concentrates on what is the most efficient and effective way to target project resources. Some questions to consider are:

- Are there people or groups whose main livelihood strategy is crop production? To what degree are they now capable of crop production?
- Are there people who are better able to farm or more skilled at farming? Will the food produced by these people be shared by everyone?
- Will there be difficulties if resources are provided only to the target groups? Are there ways of ensuring that everyone will benefit?

The second approach takes efficiency into account but looks for a way to use resources more strategically. Some questions to consider when adopting a strategic approach to targeting are:

- Are there people within the group/community who are more vulnerable and who lack power and resources?
- By receiving seeds and tools will these people benefit because having these resources will strengthen their position in the family/group/

community; or because they can trade some resources and hire labour to help them grow their own crops; or they can trade these resources in return for food or part of the harvest?

The strategic targeting of resources can bring about positive changes, strengthen traditional structures, and strengthen the capacity of people to cope with a crisis.

Some examples of strategic targeting

- Seeds and tools given to farmers' co-operatives with men and women members involving a pay-back scheme where each member returns 10 Kg of food harvested. Good farmers produce food for their families and surplus for trade. The pay-back food goes to vulnerable groups unable to cultivate. Co-operatives become functional again. More women are included in co-operatives. Some women heads of households receive seeds and tools. Agriculture extension work can be done through agriculture co-operative groups.
- Seeds and tools given to women heads of households registered for food rations. Registration cards for food can be used for distribution. Households headed by women are covered by the distribution. Women are less likely to sell these resources. Seed distributions can follow food distributions to reduce the likelihood of seed being consumed. Women are able to grow food for their families.
- Women household heads receive seeds and tools through women's groups. The women's groups do the targeting and distribution. The groups are strengthened by taking on this role and responsibility for resource management. Extension work can be done through the groups. Project monitoring can be done with these groups.

Distribution plans

At this stage, the distribution plan may have to be quite general. However, accurate figures for the target groups, target areas, number of beneficiaries, and quantities of seeds and tools required should be given, based on information from the assessment. Quantities of seed and tools should be determined according to family needs, seeding rates, access to land, and what is practical and possible, in the circumstances of funding and logistics.

Distribution plans will change as more is learned about the situation. While preparing for project implementation there will be the opportunity to include local people in drawing up the final detailed distribution plan, including a schedule of distribution times by locations, distribution sites, and target groups. This document can later be attached to the project plan as a supporting document. (See Chapter 8 for more on the final distribution plan.)

The more closely the target group is involved in formulating the distribution plan the better. There are usually people, a high percentage of whom are likely to be women, within any group whom the others trust and respect as being very fair. In the past such people may have been responsible for dividing up certain communal resources, such as meat and fish, and ensuring that the poorest are taken care of. In some situations the community can be encouraged to elect a small committee who will take on the task of choosing the individual beneficiaries. Involving responsible local officials, leaders, and co-ordinators in helping to oversee this process will also facilitate the work and maintain good relations.

Involving local structures in the distribution of inputs may be very beneficial, but not in all circumstances. Overwhelming a group with work that they are not yet capable of handling can be very demoralising and destructive. Some questions to ask about distribution plans:

- Are there existing local groups such as a co-operative or women's group? Is this group willing and able to carry out a distribution?
- Alternatively, will new structures need to be set up for distribution?
- What support will be required?
- Will this group be accepted by the community?

Monitoring

Monitoring takes place throughout the implementation phase of the project (see the project cycle in Chapter 2). Monitoring involves the regular gathering of information that is relevant to the project aim and objectives. It is essential for good management as it provides information based on experience for learning and better understanding; it indicates the level of success and the problems encountered; and it provides up-to-date information on changes in the situation which might make it necessary to modify the project.

A general description of monitoring activities should be included in the project proposal, and how the degree of success in terms of each objective will be measured. The indicators already listed identify what exactly will be measured. Monitoring needs to be planned for at this early stage and any expenses included in the proposal budget.

Evaluation

Evaluation of the project at completion by an independent evaluator may be required by some organisations and donors, and the plans and objectives for an evaluation should be included in the project plan. These can later be referred to when it is time to develop the terms of reference for the evaluation.

Evaluations are important for the same reasons that monitoring is important. They are an essential management tool and will benefit an organisation, a programme and a community in the long term. Strong reasons for including an evaluation in the project design are that there will be a lot to learn from project; the project is likely to be very difficult; or new methods and approaches to the work are being tried out. If this is a new area of work for your organisation or team then an evaluation will be very enlightening.

Project management

The way in which the project will be managed and by whom, who reports to each manager, who is responsible for what, and the lines of communication, need to be specified in the project plan. It should be made clear who will manage the budget. Accounting for the seeds and tools distributions will require records to be kept for all of the reports. Stating who will be responsible for record-keeping, and how often reports will be made, is essential.

The resources required can be listed in two or three subsections including staff required for the project, their titles, responsibilities and who they report to (full job descriptions can be added as supporting documents if necessary). All other resources required (vehicles, trucks, vehicle hire, office rental, storage rental, bags, stationary, petrol, radios, hand sets) can be itemised in this section, although they will be listed again in the budget with their cost. Relief items can also be listed with brief descriptions and quantities. Transportation may be a separate section depending on the scale of the project and the resources and funding required; when there are special requirements that need some explanation and justification then it is important to lay out a detailed transportation plan and schedule.

Project schedules or calendars can be drawn up to show when activities like purchasing, transporting, and distribution of seeds and tools will take place. They can be used to show staff activities and movements and vehicle use. They are particularly useful in complex, large-scale projects.

Budget

Your organisation and donor will have a standard budget format. (See Appendix 5 for a layout of a project budget.) To draw up a budget will require accurate figures for the cost of each item, estimates for salaries and additional staff costs like travel allowances, and estimates for the cost of consumables such as fuel, oil and office stationary. The team may decide to draw up a rough draft budget at the beginning of the planning process, which can then be modified as up-dated prices and costs are obtained. It could very well be the last thing that is completed in the written plan.

Re-checking your plans and the situation

In some circumstances the situation may change quickly. If significant changes have occurred while the process of designing and writing the project plan has been taking place then last-minute modifications to the plan may be required. It is far better and easier to make changes at this stage than after the project has been approved.

7 | Technical information on seeds and tools

This chapter provides technical information on seeds and tools. It covers the criteria and requirements for good seed, and how to maintain its quality; farmers' specific criteria and requirements for seed; what is usually available and where; the use of other propagation materials; and good practices when purchasing, transporting, and storing seed or other propagation materials. Both seeds and plant propagation materials will be collectively referred to as 'plant material' in this chapter.

Seeds

Seeds, like other vegetative materials used to grow new plants, are living organisms. And like any living organism, they require air, water, energy or food, light, warmth, and protection from physical damage. There is usually an optimum range for each requirement to sustain healthy plant life, and conditions above or below this range may cause deterioration and eventual death of the material. Maintaining the ideal storage and transport conditions for seed and propagation materials is of the utmost importance.

In general there are two categories of seed: dry seed and moist seed; in most cases it will be dry seed types such as grains, beans, and vegetables, which are distributed. Some nuts from trees are high in moisture, (recalcitrant seeds) and must remain so, and can not remain dormant for long before germination.

Moisture levels, both inside the seed and in the immediate environment, greatly affect the life of all seeds. Seeds will remain dormant (relatively inactive) until there is sufficient moisture available for plant growth; but to much moisture can exclude air and suffocate the seed. The moisture content of seed can range from as low as 5 per cent to as high as 12.5 per cent for storage.

Like all living things, seeds need to breathe; they require oxygen to create energy to maintain life, from their supply of stored plant food. They also give off other gases, that can become toxic if their levels build up in or around the seeds.

Seeds need a supply of energy to live. Seeds contain reserves of food material that provide this energy as required. The rate of metabolism (life-activity) of the seed determines the rate this stored food is consumed. Ambient or outside temperature affects the rate of metabolism; the cooler it is the slower the rate. High temperatures will destroy seeds because food stored within the seeds is used too quickly and toxic gases build up rapidly. Very high temperatures will kill off the living cells in seeds.

Light is required for photosynthesis (the process whereby green plants turn water and atmospheric carbon dioxide into carbohydrates) and is therefore essential for life and growth in most plants. Seed germination quite often requires some light to break seed dormancy and stimulate growth. The germ, or embryo, of a seed contains the plant cells that will develop into the new plant. They are genetically coded to recreate a plant that has the same characteristics as the parent plants. Increased and consistent levels of moisture, air, and light, and a particular temperature-range, are necessary conditions for germination.

The seed coat provides physical protection that helps to prevent damage to the seed from bacteria, viruses, and fungi which cause plant diseases, and from insects. It acts as a barrier to regulate the internal environment of the seed. Damage to the seed coat leaves the seed very vulnerable. Even healthy seeds are susceptible to insect attack and diseases. Infection and insect damage can take place during storage or transportation, and when the seed is planted. Seeds are often coated with insecticides, fungicides, and other chemicals, to protect them. This can be done by the seed producer, the agent selling the seed or by the organisation using it. In relief projects there is always a risk that some people will consume the seed as food. There is, therefore, a possibility that people might be poisoned by eating seed treated with chemicals. It is therefore advisable to use only untreated seed for distribution in relief projects.

Plant groups

Plants are classified into families, genera, species, and varieties. (See Appendix 6 for a full list of common crops with their Latin names.) Families are a very broad grouping, for example, Gramineae or grasses. Genera are more specific, for example, sorghum. Within each genus (singular) there are species, such as bicolour. Species cannot be interbred because the genetic differences between them are too great. Each species contains many varieties, and with a few exceptions, these can be interbred.

The extremely different and varied characteristics of each crop or species, such as rice, groundnuts, millet, potatoes, maize, bananas, and cassava, and the more subtle differences between varieties of a single crop or species, are governed by genes, the genetic material contained within every living cell. Genes are located on chromosomes, strands of genetic material in the nucleus of the cell. They are paired, with one of every pair having come from each parent. When cells divide in normal processes of growth, the chromosomes reproduce themselves exactly, and each new cell has a full complement of identical chromosomes. When reproductive cells are formed, the process is quite different: the chromosome pairs split up, and one of each pair goes to each reproductive cell, or germ cell. During the process of chromosome division, it is possible for the genetic material originating from each parent to be recombined, so that the newly-formed germ cells may have a different assortment of genes on the chromosomes, with genes having come from either parent. Sexual reproduction, in animals or plants, involves the joining together of germ cells from the male and female parent, so that the resulting embryo has the full complement of genes. Seeds therefore contain cells with all the genes necessary to reproduce plants exactly like the parent plants.

Propagation using other parts of the plant

For some crops, new plants can be produced without seeds by using parts of the plant; for example, pieces of roots or tubers of crops such as cassava, potatoes, and yams. Pieces of stem or cuttings can be used for cassava, tomatoes, and many fruit trees. More complete plant parts called shoots and suckers are used to start new plants in the case of bananas and pineapples. This process of propagation without the use of seed is termed vegetative reproduction. Each new plant is exactly like the parent plant, because no changes in genetic make-up can occur, as they can through cross-breeding. Vegetative reproductive methods are particularly useful for crops that do not produce seed easily, abundantly, often enough or true to variety.

Plant materials are more difficult to store than seed. They can only be stored if the conditions are exactly right, and even then, not for long periods. Some roots and tubers can be stored for a while in the ground where they have grown. Most plant propagation material is freshly prepared for planting, and used as soon as possible. Roots, stems, shoots and suckers are bulky compared to seeds; this makes handling these materials problematic, especially if large quantities are required. However, when necessary, it is possible to store and distribute such material successfully. The conditions for storage and transport are similar to those for seeds.

Plant breeding

Seeds are the only plant material that can have a new genetic make-up as a result of cross-breeding between two parent plants. In cross-breeding genetic material is mixed naturally to create new combinations of genes within chromosomes, which in turn produce plants with some characteristics of each of the original parent plants. Plant breeding involves selecting varieties with desirable characteristics, inbreeding them for genetic purity, and finally cross-breeding two varieties to produce a new variety or hybrid. Sometimes it requires several crosses, with three or four different parent varieties used to achieve the desired end-result: a new hybrid.

Some seeds will continue to reproduce plants with exactly the same characteristics, generation after generation, while others will not. Hybrid varieties are genetically unstable and only retain what is called 'hybrid vigour' from one to a few generations. By the time they are multiplied and sold in quantity they are only capable of producing one stable homogeneous population of plants. Saving seed, to grow another crop in the next season, may prove disappointing, producing a mixture of different plants, some of which will be weak and unproductive.

The advantage of these hybrid varieties is their high yields. Scientists have focused their breeding programme for a particular crop on specific qualities such as production of seed, in the case of food-grains. Some varieties are also bred to be resistant to certain diseases. Unfortunately, concentrating on specific characteristics means that others are lost. For this reason many hybrid plants are more susceptible to diseases, and damage from birds and insects. They are not as hardy as the old standard varieties; they are not as good at taking nutrients from the soil and at competing with other plants for space, water, and nutrients. Hybrid plants need more care and protection, and require inputs such as fertilisers, insecticides, fungicides, and herbicides if they are to succeed.

Standard or traditional non-hybrid varieties are much more stable genetically. Their seeds will produce plants that are true to type. In normal reproductive processes, there is the possibility for some genetic mixing in most species, so varieties will naturally occur. By a gradual process of natural selection, they will slowly adapt to changing conditions or migrate and adapt to new conditions. Humans have learned to assist and guide this process. For thousands of years farmers have been harvesting seeds from selected plants that have the qualities they prefer. They have also moved seed and other plant materials around to find new niches for the crop. This selective process is responsible for many standard (improved) domestic varieties of crops that are grown today. Grains have been selectively evolved over millennia from their relatives, the grasses (the similarities are obvious: the long narrow leaves and characteristics of the seeds).

Of these standard open-pollinated varieties of crops, also known as 'local landraces', there are some that cross-breed or cross-pollinate easily, while others have a flower structure or shape that inhibits this. Some crops, for example, apple trees, only produce viable seed when cross-pollinated by a different variety; while others, such as pawpaws, have separate male and female plants. Crops that are prone to self-pollination remain true to type, making seed collection and use easier; examples are oats, barley, and tomatoes. Varieties of crops that readily cross-pollinate, for example, maize, sorghum, and millet, must be isolated from each other geographically to prevent cross-breeding and so produce pure seed. Besides physical distance, natural barriers like forests and other crops (species) are used to prevent cross-pollination between varieties. Varieties of crops which have different flowering periods, because of slower growth to maturity or different planting times, cannot cross-pollinate easily.

Vigorous plants that have all or most of the desired characteristics will provide the best quality seed that in turn will produce the best crop the following season. When a whole crop is grown for seed, weak or different-looking plants are removed just before flowering and again just before harvest. Both good farmers and seed producers follow these practices, which help to maintain the purity of a variety.

Selecting seeds for distribution

Choosing seed for farmers to grow should involve the local experts, the farmers, both women and men. It is increasingly recognised that women often have extensive knowledge of seed and plant varieties. For example, an NGO working in Liberia providing rice seed to farmers after a period of conflict experienced problems because the imported seed to be distributed was a mixture of highland and lowland types of rice which were not clearly labelled. The problem was solved when local women farmers inspected the seed and identified the different types so it could be separated, marked, and distributed appropriately.

Farmers will usually produce a list of diverse crops (species) and varieties required by them to produce food and cash-crops on their land, with its specific micro (small-scale) and macro (large-scale) environmental and climatic conditions. Each variety will fit into a special niche in terms of soil type, soil fertility and moisture levels; cropping systems such as inter-cropping, crop rotations, companion planting, staggered planting, and seed production; domestic requirements such as early food crops, late crops for storage, grain for oil production, grain for brewing beer, grain for making flour, and crops that can easily be sold or traded for other items. Farmers' needs are very specific and they require varieties of crops that they know well and can depend on.

Agricultural extension work

Farmers usually experiment with new varieties and crops on a scale and at a pace that involves very little risk. They can not afford to make big mistakes which could affect their livelihoods. Government and NGO agricultural extension workers have to counter this conservative approach when they try to introduce new crops, varieties, and practices. When farmers and advisors work closely together with mutual respect the results can be very beneficial. Quite often agricultural extension officers work with particular individuals who are willing to try out whatever is being introduced. If it works well for them, then other farmers may try it.

However, there is a potential danger associated with extension work in agriculture. Farmers may be persuaded to adopt a cropping system or introduce a new variety which completely fails in a bad year. Farmers may become more dependent on cash-crops for their livelihood, that depend on very unreliable markets. Monoculture crops, where only one crop and often only one variety is grown, are very susceptible to insect and bird infestations, in contrast to traditional systems of agriculture where a wider range of food crops and different varieties of a crop are grown.

Some common food crops

Seeds for the following crops are commonly distributed in relief projects; all of the numerous varieties of each species are not named here for obvious reasons. It would be useful to compile regional lists of varieties with detailed descriptions.

Grains

Grains always seem to be a top priority, because they are often the staple food. They are regarded as important, even by agro-pastoralists. Grains can be harvested and stored for relatively long periods of time under the right conditions. They are high in both protein and calories, and are adaptable to many different methods of preparation. They are easy to transport and to market because of their high value-to-weight ratio. They are propagated by seed, which is easy to gather, store, transport, trade, and sow. There are local methods used to treat seed, which are much safer and more environmentally friendly than using commercial fungicides and pesticides.

Sorghum: One of the most widely used grain crops in Africa, sorghum has hundreds of varieties with a wide range of characteristics adapted to many different growing conditions. Sorghum is a traditional crop in many places, with local standard varieties commonly preferred and used. These standard varieties are open-pollinated and can be used for seed but some varieties easily cross-pollinate. Plant breeders have produced sorghum hybrids. These are considered better commercial varieties because they produce higher yields under the right conditions.

There is a growing tendency for farmers to reduce the amount of sorghum they grow and replace it with hybrid maize that is higher yielding and easier to market. Sorghum, however, is generally less susceptible to dry conditions and therefore much more reliable. Some sorghum has a very short growing season and is used as an early crop or for two consecutive crops in one year. There are mid-term varieties that can be harvested after the short-term ones. The late varieties are important for storage but are more prone to failure when rains taper off early in the season. Some late varieties have a high sugar content in their stems (like sugar cane) which can be eaten raw. Imported varieties of sorghum may be adapted to an area with a different day-length, and so may not produce seed at the right time, or at all. Sorghum grain has quite a high protein content (10 per cent) but this protein is not adequate for a balanced diet, so sorghum needs to be used in conjunction with other grains like rice. (The protein content of some common foods is given in Appendix 6.)

Millet: Another traditional crop that is widely used in dry conditions in Africa. Farmers use standard varieties of bulrush millet and finger millet. Inter-cropping millet with sorghum is a common practice where seasonal rains are not reliable. Millet is not a popular cash crop although the grain is often bought and sold in local markets. It has the same amount of protein as sorghum. Its seed is not usually distributed in relief projects and may be difficult to find and purchase in large quantities.

Maize: Maize is being introduced into many areas as a high-yielding cash-crop. Most varieties available are hybrids so new seed must be purchased by farmers every year. Maize is not drought-resistant as it is shallow-rooted. Its requirements for good moisture levels, fertile soil, and pest control make it a demanding crop. High yields often depend on inputs like chemical fertilisers, weed-control, and insecticides, adding to the production costs. Maize has been used in seeds and tools relief projects with varying success. Along fertile moist river banks or where irrigation is possible it can be grown well even out of season.

Rice: A traditional staple crop grown all over the world, and successfully introduced into new locations including regions of Africa. Highland rice, as its name implies, is adapted to moist but well-drained soils usually on sloping terrain. Lowland rice is adapted to swampy land that is naturally flooded or has been developed into rice paddies. There are thousands of different varieties of rice; many have developed locally through naturally occurring variations being selected by farmers, and some are adapted to very specific environmental niches. Rice seed can be saved and used by farmers. Improved varieties are available commercially in large quantities.

Root crops

Root crops are very important to farmers but only too often they are not considered for distribution in relief projects because they are propagated with bulky plant material such as roots and stems. This makes transport and storage more difficult and expensive; for this reason, it is difficult to find sources of plant materials in large quantities. Much of it is produced locally. However, root crops are very valuable because of their special characteristics. They are adaptable to a wide range of environmental conditions. They are flexible in terms of harvest time and store well in the ground making them an excellent drought and 'hunger-period' food. Including them in a distribution project, even in small amounts, can improve food security in the short term and ensure food security through hungry periods in the long term.

Cassava: A traditional root crop that is widely used and well adapted for cultivation in dry regions. Standard varieties are often used to supplement staple crops. These can take two years to mature. Cassava is very important as a dry season (hunger period) and drought food because the roots can remain in the soil and be dug up at any time. The leaves, which are high in protein, are also eaten. Varieties with a shorter growing season are available but they require careful preparation (in quantities of water) before they are safe for consumption. Cassava is often found untouched and still healthy in the ground after periods of drought or insecurity. During or after a crisis cassava roots and leaves may be the only food available.

Cassava has not regularly been used in relief projects. This may be due to the fact that propagation must be done with stem cuttings, which are difficult to organise and distribute. It seems that the importance of cassava is not fully recognised by relief workers.

Unfortunately cassava is susceptible to a disease caused by a virus called cassava mosaic. It is spread from plant to plant by the white-fly, a tiny pest that feeds on the leaves, and propagation material may carry the disease. Pale-green blotches and deformed leaves are two obvious symptoms. Lower yields and weakened plants are the result. There are disease-tolerant varieties available, which are not resistant to the disease but will tolerate it to some degree. Quality control should include securing disease-free parent materials for cuttings. Control of cassava mosaic is a problem that can only be dealt with by a long-term development programme, involving crop breeding, vector control, and improved farming practices.

Legumes

Most legumes have large seeds and some (like groundnuts) are best transported in their shells. This makes them more bulky and so more expensive to transport. Like root-crops they have special characteristics and should be considered for certain situations. Farmers know from experience that crop diversity is important especially where weather conditions vary from year to year and may be extreme in some years. Providing a range of crops that local farmers consider important will help them to fight against the ill-effects of pests and diseases, drought, or flooding.

Legumes are high in protein. They often have edible leaves. There are legume varieties which are adapted to different environmental conditions; and they have the added advantage of actually increasing the soil fertility because they can turn atmospheric nitrogen into nitrogen compounds in the soil available to plants. They carry nodules on their roots in which nitrogen-fixing bacteria live, in a symbiotic relationship with the plant in which both parties benefit. For this reason farmers use them in crop rotations with grains and root crops to maintain soil fertility. Where the specific legume crop has not been cultivated before it may be necessary to inoculate the soil with specific nitrogen-fixing bacteria that live in the root nodules of that crop.

Groundnuts: A traditional crop widely used and very popular because of its high protein content and ability to fix nitrogen from the atmosphere. Standard varieties are used by many farmers and are favoured for their oil content and taste. Hybrid varieties are available with larger nuts and higher yields. These have the usual demanding requirements of hybrids, making them somewhat less suitable in difficult conditions. Groundnuts are used in relief projects but seldom supplied in quantities that meet the farmers' full needs. The seed travels and is stored best in the shell. Farmers use groundnuts in crop rotations because they improve soil fertility. Groundnuts require the presence of a symbiotic soil bacteria for best results.

Beans: Another legume that is widely used and adapted to many conditions. Like groundnuts, beans are popular because of their high protein content and their ability to fix nitrogen, and are used in crop rotations to improve soil fertility. They require the presence of nitrogen-fixing soil bacteria for best results. These are not always present in soils where beans have not been grown before. Farmers can use standard varieties to produce seed.

Vegetables

Although they are not high in protein vegetables are very important complementary food crops. They provide vitamins and minerals necessary for a balanced, healthy diet. They also provide taste, texture, and colour to increase the palatability of staple foods. Green vegetables tend to be the highest in general food value. Vegetables are usually grown around the home in small plots, making it easier to care for and protect them. Many vegetables have the added advantage that it is possible to grow them in the dry season if water and some shade are available. Vegetables are an important cash-crop. Women commonly grow and market vegetables to earn extra income for the household. They seem to be easy to market in most places.

For the average family garden six to eight different types of vegetables, with seed weighing from 15 gm in total, is sufficient. (These requirements will be higher for market-gardening and for crops like beans and corn that have bigger seeds.) Standard varieties of vegetables can be used to produce seed in some cases. Vegetable seed varies greatly in its storage requirements. Some seeds (e.g. carrots) do not last well beyond one year; the germination rate drops drastically unless special storage conditions are provided. A minimum level of basic practical knowledge is required to grow vegetables. When this is lacking, extension work needs to be provided along with the seeds to avoid failure and disappointment.

Tomatoes: There are many hybrid tomato varieties available (and in fact it is becoming increasingly difficult to find standard varieties). Tests have been shown that often the increase in size of hybrid plants (and fruit) grown with chemical fertilisers is simply due to a higher water content in the plants. Many gardeners say that standard varieties produce fruit that tastes better and stores longer. It is quite easy to save seed from standard tomato varieties.

Eggplant (aubergines): are similar to tomatoes in many ways and are higher in protein than most vegetables (4 per cent).

Cucumbers: It can be difficult to obtain varieties that are useful for seed production. There is very little food value in cucumbers.

Cabbage: This is popular in some areas. The seed is difficult to save under normal conditions. Cabbage can be a good cash-crop.

Peppers: Sweet peppers and hot peppers are similar to tomatoes. With standard varieties seed can easily be saved. They are a popular cash-crop in many places.

Bush beans: a legume, they are very high in protein and improve soil fertility when used in rotation with other crops. Standard varieties can be used by farmers to produce seed.

Peas: Are much the same as beans but are not as tolerant to heat.

Okra: A traditional crop widely adapted to conditions in many places. Seeds are saved from the standard and local varieties available. They are a very good local cash-crop.

Carrots: Not as tolerant to hot dry conditions as other vegetables. The seed is very difficult to produce locally.

Onions: Quite popular in many places; a good cash-crop for local markets. With some skill they can be grown under a range of different conditions.

Lettuce: Not very tolerant to heat, sun, and dry conditions; it bolts or heads quickly and then becomes bitter. There are often local plants used for green leaves that are much more suited to hot climates.

Sweet corn: Does well with a lot of sun but needs adequate soil moisture and fertility. Seed can be saved when standard varieties are used.

Tools

Hand tools have been described as 'an extension of the human body'. They make work easier by providing some mechanical advantage, such as leverage; by providing a sharp strong edge with which to cut; by providing a strong flat surface to pound and crush; and by allowing the person to work in the most comfortable position. Tools assist people to do their work faster, more efficiently, more effectively, and with much greater ease.

Tool design

The shape of a tool and the strength of the materials used to make it are very significant. Farmers have many tasks they must perform in order to produce food, and the tools they use are very important to them. Because tools break, wear out or are lost, farmers need a reliable source for new ones. Some farmers can make their own tools, but most farmers rely on local blacksmiths to manufacture what they need or they purchase tools that have been mass-produced. Farmers need tools that are well suited not only to the specific task but to their own physical characteristics.

Whatever the type or source of tool used it has been found that most farmers ask a blacksmith to modify it to meet their specific individual requirements, which depend on the soil types, the kind of cultivation done, their physical characteristics and strength, and their methods of working. For example, an elderly women may need a lighter hoe than a young male farmer. Women may want a smaller hoe to be used specifically for weeding or a farmer may want a very strong hoe to cultivate heavy clay soils. Mass-produced tools will not suit everyone.

The metal used for tools must be carefully specified. Carbon steel can only be worked at temperatures of 450 degrees, which is not achievable in local traditional workshops. When tools will be modified, carbon steel should be avoided, as tools made with this tend to break when being worked by local blacksmiths. A workable steel suited for local blacksmiths is mild steel that has not been hardened. It is softer and therefore not as ridged so it can easily be forged. Blacksmiths can give tools made of mild steel strength by working them and modifying the design. Whether tools are being purchased abroad or regionally, it is good practice to have them tested if there are testing facilities available. Tests can be conducted by independent metallurgists for a fee. They will analyse samples to determine if the tools are reasonably fit for their intended purpose. The chemical analysis shows the carbon, sulphur, silicon, phosphorous, and manganese contents; these elements affect the characteristics of the metal.

Tool manufacturing

Tools are mass-produced by casting, machining or stamping them out. The stamping process is most common because it is very fast and uniform but it does have limitations. Local blacksmiths forge tools, a traditional process which is slow and produces less uniform tools; when the thickness of the metal of a tool must vary to give it strength in one area, keep it light or allow a handle to be fitted locally, the tool must be forged.

Local blacksmiths should be involved in tool selection from the beginning. They will be able to tell your team about tool preferences, sources, the best metals for certain tools, and the modifications usually made locally. They may be put out of business unless a way to include them is found. Employing local blacksmiths, perhaps collectively, to manufacture a very specific tool like small hand-hoes, or to make a proportion of the required tools, or modify tools for certain groups of people may help to rebuild their capacity to provide their services to the community or group in the long term.

Tool handles are best fitted by the farmer or local blacksmith. Unless there are no local materials, such as bamboo or small timber, to fashion handles from, it is not necessary to distribute handles along with tools. They are traditionally cut, modified, and fitted to the tool by each individual so that they meet exact personal requirements. This allows the user to work more efficiently and effectively with less physical strain.

8 | Project implementation

This chapter will cover the specific activities involved in implementing a seeds and tools distribution: purchasing the items for distribution, storage and transport, the actual process of distribution, monitoring and evaluation, and writing the final project report.

The purchase of seeds and other propagation materials

The crop varieties to be distributed should be limited to two or three types, agreed upon by the beneficiaries. It may not be possible to obtain the whole quantity of seed required from a single source, and some varieties may not be available at all. Several suppliers and alternative varieties may have to be considered. Large suppliers expect to include transport and seed treatment as part of the contract; these additional costs should be allowed for when making comparison with costings from small local suppliers. Some suppliers consider seed treatment standard practice; it should be made clear to them if this is not desirable. The advantages and disadvantages of various sources need to be considered carefully.

The advantages of local seed sources, when and where they exist, are that they will be able to supply varieties that are well known and suited to local conditions. Propagation materials can be supplied fresh. There is no risk of introducing diseases and weeds new to the area, and less chance of delay in delivery. Direct quality control will be easier. There will be very little if any bureaucratic red tape or transport problems, and the costs of seed and transport costs are likely to be lower than from other sources. Purchasing locally may also stimulate the local economy.

The disadvantages of purchasing locally are that it can demand a great deal of organisation especially when many small suppliers are involved, when it may be necessary to negotiate several separate contracts. The quality of local seed may be very inconsistent. Your organisation may have to take responsibility for cleaning, packaging, labelling and transport of seed and materials, and for quality control. Sufficient quantities may not be available, and the effect that purchasing seed locally may have on the local market and farmers must be considered carefully; in some cases, if there is a general shortage of seed, purchasing locally may do more harm than good.

Some possible sources for local seed and propagation materials are:

- Traders, who may not be used to dealing in seed or large quantities of seed and who may sell food grain as seed, which may not have been stored in ideal conditions.
- Local agents or traders who have a good reputation for organising such transactions and who know the local market.
- Farmers' groups or co-operatives that have surplus seed and may or may not need resources and support to organise transport, packaging, and storage.
- Government or NGO agricultural projects that may have the capacity to supply seed and propagation materials but may need resources and support as above.
- Companies that contract skilled farmers to produce seed or propagation materials for them; these may or may not be reliable and documents such as seed certificates and contracts will not always guarantee quality and standards.
- Local markets may have small quantities of seed for sale.

Large, centralised suppliers of seed and propagation materials, when and where they (still) exist, should be able to provide varieties adapted to the conditions in the project area. Commercial varieties, often hybrids, might be the only seed available in any quantity because government programmes and economic pressures greatly influence these larger markets. Large suppliers often organise transport, cleaning, treating, packaging, and labelling. Large suppliers may charge high prices but offer seed that is certified to be of a high standard. The extra cost may be justified: poor quality seed wastes farmers' precious time and resources.

However, there are some disadvantages. The varieties on offer from large central suppliers may not be what the farmers really want. Hybrid varieties of seed can not be saved by the farmers for sowing in the following season, and they may require more care and inputs like irrigation, fertilisers, and pesticides.

It may only be possible to obtain the required seed from sources outside the country, because of the internal security situation or simply because of unavailability within the country. Suppliers large enough to deal with international orders will often be the big central companies in other countries. It will be more difficult to find varieties that are adapted to the project area and much more difficult to find varieties that are the same or similar to the ones familiar to local farmers. Communication with the supplier will be more difficult, increasing the probability that something will go wrong. Quality control tests may have to wait until the seed arrives.

There are many problems that can arise between NGOs and seed suppliers. When dealing with any supplier is on a 'one off' basis, there is less reason for them to give good service. When the clients are inexperienced or in a big hurry they may be taken advantage of. A golden rule is know as much as possible about the crops and varieties or alternative varieties required and never, ever purchase anything out of desperation. If in doubt, stop; poor seed is often much worse than none at all. Late arrival of seed is the most common problem. Seed may have poor germination rates, and may not be consistently of the high standards of the samples originally provided. Seed may be labelled as all one variety, but turn out to be several different varieties, and not those required. Or an alternative variety may be said to be adapted to conditions in the project area, but in fact perform very poorly. Potential suppliers should be vetted by as many reliable sources as possible: other NGOs, agencies, the Ministry of Agriculture, local agriculturists, and local women and men farmers. Suppliers should be asked about their quality control methods, sources, guarantees, and other customers, and their facilities and seed crops inspected if possible.

The contract with the suppliers is very important. A reputable supplier will only give contractual guarantees that they have a reasonable hope of meeting. A great deal can go wrong, and a supplier who talks openly about potential problems, and who carefully negotiates the contract, will probably take equal care in keeping the agreement.

Quality control and standard tests

Whether quality control is all or partly done by the purchaser, it is essential for the success of the project. There are standard tests and criteria for seed quality, and the results of these tests are indicated on seed labels and certificates provided by the growers and suppliers. They are valid for that seed at that time of testing so batch numbers and testing dates are significant. The value of government certificates issued depends on the testing standards of that particular agricultural department. The value of commercial claims and labelling as to the seed quality depend completely on the supplier's integrity. Seed and propagation materials that are of high quality when purchased will quickly deteriorate if not stored, handled, and transported with care.

Testing should be carried out by reliable, independent (i.e. independent of the supplier) experts or qualified staff. Seeds should be tested before purchase, upon arrival into stores, and again just before the actual distributions. The second tests are to verify the first and check for rates of deterioration. The third test is mainly to check on rates of germination before giving the seed to farmers, and can easily be done by the team. Most tests are not difficult to perform but require an understanding of methods of sampling and estimating percentages. Seed are usually tested for:

- the germination rate
- the percentage of weed seeds present
- the percentage of damaged seed
- the percentage of other crop seeds present
- the percentage of other materials (dirt) present
- the percentage purity of variety (this is a more difficult test)
- the quality of seed by size or weight per 100 seeds

When time is short you will have to rely on quick germination tests and observation of samples of the seed. Local farmers are often experts on seed characteristics and variety and may be able to help in assessing seed quality.

The purchase of tools

The specific requirements for each type of tool will determine the manufacturer or source. Some tools, for example, axes, shovels, picks, and rakes, are fairly standard; there may be slight differences in the types farmers prefer but their biggest concern will be good quality. Finding a source for these tools should not be a problem; there may be a standard make that is popular because it fits most requirements. The best design details have been worked out over the years, and quality is usually determined by the standards of metal used. Experienced farmers will tell you the brand names they prefer. Price is often one indicator of quality and poor tools have very little or no value.

However, there are other tools, such as hoes and machetes or pangas, that farmers use a lot for which there may be local specifications. When farmers ask for a very specific design of tool, whether it is similar to or completely different from a standard model, it may be much more difficult to find a manufacturer who can produce these tools in the required quantities. With tool designs that are not standard or manufactured on a large scale regionally or internationally, you will have to be involved in design specifications and quality control from the beginning.

The most common sources for tools are:

- Local blacksmiths, who know the specific needs of farmers but can only produce small quantities of tools. In some situations, particularly where there is or has been conflict or insecurity, they may no longer have the metal or tools they need.
- In-country manufacturers, large and small, who already turn out standard popular designs of tools. They may or may not have all of the resources they need to produce the tools required.
- Manufacturers based in other countries who export their products internationally will have standard designs that are widely acceptable. They can usually provide everything needed in any quantity, but their standards vary greatly. Transport of tools over long distances by air is expensive; by sea it costs less but can be very slow.

Purchasing procedures

Some NGOs require at least three quotations from different sources when purchasing anything. Ask each potential manufacturer for the specifications of their range of tools, samples of each when possible, and prices. When asking for estimates for the price of tools that are not of a standard design you will have to supply each manufacturer with the specifications and samples. Often the more you order the cheaper the unit cost.

Many organisations have purchasing departments that have been set up to assist projects. They have knowledge and experience of international suppliers, and usually have reliable information on sources, prices, quality, transport, and import/export procedures. They may also advise that the best possible source for the tools required is a local one.

Local blacksmiths and farmers can help to select the best tools from the samples received from the various sources, or from specifications, drawings, photographs, and descriptions. Remember that women will have a lot of constructive input into this process as they do much of the farm work and have specific requirements for tools. When specific local designs of tools are required it is especially important to involve local farmers. If samples of modified or specifically designed tools are sent by manufacturers, they can select the most appropriate ones.

Quality control

Manufacturers are usually controlled by a National Bureau of Standards of the country they are based in. They often state in their codes that 'for the sale of goods by sample the bulk of the goods must correspond with the sample in quality'. Nationally or internationally recognised quality inspection agencies can be used

Transport and packaging

Transport of plant materials

Transport of seed and other plant materials should be thought of as a mobile stage of storage, during which optimum conditions should be maintained. Plant materials should be kept reasonably dry. Excess moisture can be damaging by suffocating the material, starting germination or growth prematurely, encouraging mould or fungus growth which can cause rotting, and by contributing to overheating. If seed or propagation materials get wet they must be carefully dried before storage or used immediately. When partial germination has occurred as a result of seed getting wet, drying will kill off the seeds.

Overheating quickly shortens the life-span of seed and other plant materials. It can directly kill plant cells. It reduces the vitality of seed and plant materials so that they make poor growth and produce weak plants. Prolonged exposure to heat will greatly reduce the quantities of viable seed or plant material. Extreme cold is less likely to be a problem, except possibly during air transport, but manufacturers sending goods by air are usually aware of the effects of cold temperatures, and protect consignments accordingly.

Ventilation is important; plant material needs to breathe. When it is packed tightly in a lorry, air circulation is unavoidably reduced; therefore transport time must be kept to a minimum. A combination of heat and poor ventilation is extremely damaging, because of the build-up of toxic gases. When delays occur during transportation, it may be necessary to unload and restack all seeds or plant material.

Seed for food-grain is most commonly packed in 50 or 90 kg sacks. As the sacks will probably be handled manually they should be light enough to lift, and made from strong materials so they do not break open. (It can be very time consuming to clean up seed spillage in trucks and storerooms.) Adequate labelling avoids confusion, and is especially important when different varieties of the same crop are involved. If possible a label should be wired to each sack and another placed inside. If seed is to be air-dropped from planes it must be double- or triple-packed in strong bags to avoid breakage and spillage.

Vegetable seeds supplied in large quantities often come packed in tins. Re-packaging may be required when small amounts of vegetable seeds are being distributed. The beneficiaries may be able to divide up seed provided to them in large bags, or may need it to be packaged in much smaller bags for distribution. This must be taken into account so that the necessary resources and labour are organised; inappropriate packaging can cause confusion, chaos, and delays at the distribution sites.

Transport of tools

Some tools are very heavy and therefore can be expensive to transport. Trucks will have a maximum payload, and this limit should not be exceeded. In areas with very poor road conditions loads should be reduced below this maximum; trucks which get stuck, or break down, can cause long delays and the loss of goods. A truck loaded with tools to its maximum limit by weight may not be filled up in terms of space. It would therefore be better to have mixed loads of light and heavy goods to make up a load that fills the truck but does not exceed the weight limit.

The supplier may be able to pack the various tools you have ordered in quantities and in containers that will make unloading and reloading by hand easier. Sharp tools are better packed in boxes. When tools are packed in bags the bags should be made of material that does not rip easily. Some tools like hoes are simply wired together in bundles. Tools may be unloaded and reloaded as many as six times before they reach the farmers, so each unit or package should be strongly secured, not too heavy, safe and easy to pick up. When aircraft are used to transport tools special care must be given to packaging as sharp protruding edges can easily damage the planes' lightweight interior. Pilots will simply refuse to take a cargo if it looks poorly packed.

Storage

For safe storage of seeds and propagation materials, a warehouse or storeroom will be required which is large enough to store the quantities required, dry, secure, clean, as cool as possible, and easily accessible to trucks. It should have a sound floor, be protected from rodents, and be well-ventilated. The storage area required can be calculated on the basis of the maximum amount of grain or plant material to be kept there at any one time. (See Appendix 7 for details of how to calculate storage areas.) Plant material should be stored on pallets, if possible. Pallets are slatted wooden platforms which keep the grain bags off the floor, which may be damp, and allow air to circulate round the stack. Plastic sheets or tarpaulins can be used as an alternative if the floor is damp and pallets are not available.

Quite often repairs and improvements will be necessary to bring a warehouse up to the required standards. Tarpaulins or plastic sheeting can be used to mend a leaky roof quickly or to provide shelter for seed where no warehouse is available. Security may not be a problem but it is always wise to have locks on all doors and guards for each building.

Overheating may be a serious problem. Metal roofing radiates heat down into the room. Dried straw or grass placed in a thick layer on the roof will help to keep temperatures down. If metal containers are used they can also be protected in this way. Ideally temperatures should be kept below 50°F, but a range of 70–85°F is acceptable for short-term storage of most seeds. Try to ensure that the temperature always remains below 95°F.

Most warehouses have openings or windows to allow for ventilation. These should be screened and made safe from intruders. If air circulation is poor the doors may have to be kept open (and guarded) during the day.

Damage from insects and rodents is common; it may be necessary to use traps and poisons in the warehouse. When this is necessary great care should be taken to safeguard the health of any staff working in the warehouse. Pesticides, especially when used in confined spaces, are very dangerous. It may be better to avoid using poisonous chemicals, accept some losses, and keep storage periods very short.

The main concerns when storing tools are adequate space, good access, security, and protection from damp. It can be difficult to calculate the space required, as this depends on how the tools are packaged. Spacing can be tighter than that for seeds. (See Appendix 7 for details of how to calculate the space required for tool storage.)

It may be necessary to divide seeds and tools between more than one warehouse; perhaps putting tools in one and seed in another, using the coolest or biggest building for seed. Or it might be more convenient to store some of each in each building so that loading the trucks is easier. Items should be stored in a way which makes it easy to load them on trucks for transport and distribution. If tools will be needed first they should be stored in front of the seed.

Stock records

A simple system of recording everything that comes into the store and everything that leaves is essential. One person should be responsible for maintaining stock records, and should keep the keys to the store. Stock should only be issued with authorisation from the project manager and from the store-keeper, and records should be checked weekly. These records will be needed for reports, and for checking on stock levels during the distribution.

Distributions

The selection of crop varieties, types of tools, and the detailed distribution plan need to be decided on in close consultation with the beneficiaries. Sometimes it is appropriate to set up a committee of women and men representatives of communities or groups of beneficiaries, key informants, local leaders, extension officers, and local authority representatives. Try to ensure such groups are gender balanced, with a majority of beneficiaries (farmers). Local agriculturists and farmers may play an important advisory role, especially if the beneficiaries are not indigenous or have little experience of farming. If there are no local people available who are knowledgeable about agriculture in the area concerned, advice should be sought from agriculturists within the region, or from elsewhere in the country.

Discuss the project design with the committee, explaining any limitations or constraints. Ensure that they agree that the project:

- targets appropriate groups
- will work with the appropriate local structures or creates structures that they can relate to
- has realistic objectives
- is timely and will provide all the necessary inputs.

Plans should be sufficiently flexible to be improved on and modified, as a result of these discussions, or later, during implementation. Always take notes during these meetings as there will be much to learn. Be open and honest about project weaknesses and any mistakes made in the design, as people will respect this approach and either accept the limitations or suggest ways to overcome any problems which emerge. This meeting could be used to discuss details of distribution plans, such as who will identify individual beneficiaries; who will carry out registration and distributions; the system of distribution to be used; and the most convenient distribution sites.

Final distribution plan

A final detailed distribution plan should now be drawn up, including references to consultations and decisions made at these meetings. This plan can be used as a supporting document attached to the project plan, or part of a progress report. The elements to include in any distribution plan are as follows:

- exact target group/s
- the numbers of intended beneficiaries
- where each group are geographically located
- how each target group will be assisted, forms of support and inputs
- how each family or individual will be assisted
- who will identify and register the individuals/families to receive assistance; if this will be done by a group or committee, how it will be selected
- the support this committee will receive and from whom
- the distribution structures that will be used and support provided to them
- the distribution methods to be used
- the location of each distribution site (a map)
- description of a distribution team, responsibilities of team members
- samples of the distribution record sheet, registration card, and registration sheet to be used
- distribution schedule with places, groups, distribution teams, and dates.

Management of a distribution

Distributions provide excellent opportunities for contact with beneficiaries and the development of working relationships between them and the project staff. If possible one staff member should be free to move around and talk to the people present at the distribution site; this could be the team leader who is managing the distribution. Valuable information can be gained in this way, and problems can be discovered and quickly put right. It may be a chance to gather information on areas that are inaccessible because of poor roads or insecurity.

Registration

Carrying out a registration of individuals or families before the actual distribution takes place has many advantages. An accurate figure for the number of people who will receive the seeds and tools can be obtained. Grouping people by areas, camps, or villages will facilitate setting up a distribution schedule with specific sites and dates. Announcing these well ahead of time will allow people to plan their other activities around the distribution, especially when distances to distributions sites are great.

Registration sheets can be used to record basic information useful to the project in the future, but they should be kept as simple as possible. They can also be used to tick off distribution items received, although it is time-consuming to look through sheets to find a person's name each time. Registration cards can be given to beneficiaries, which record basic information: name, sex, number of family members, acres farmed, group or community, head of household, and items received. These cards are then presented by the beneficiaries on distribution day, speeding up the whole process. (See Appendix 8 for sample registration/distribution sheets and cards.)

Take time to explain the distribution plan to the community leaders and to the field staff and volunteers who will be involved. Clear, careful explanations, and practice sessions with the field staff, will help to ensure that the correct procedures are used and the actual distribution goes smoothly. Misunderstandings occurring during distributions can be very disruptive and difficult to resolve.

Tasks of the distribution team

The are four separate tasks for the distribution team for a seeds and tools project. The team leader has the task of delegating responsibilities, ensuring that all the other tasks are being done properly, ensuring that everyone on the team has the necessary support and resources, and that problems arising are dealt with. She or he can also make contact with the local leaders and authorities, monitor the distribution, help with tasks when required, and talk to local people. The other tasks are:

- Checking cards and recording information, such as names, villages, numbers in household, and ticking off items received. Records and record keeping should be carefully planned and organised; these records will be crucial for project reports and future planning.
- Issuing the seeds and tools to the beneficiaries after their names have been taken.
- Organising people at the distribution so that it takes place in an orderly fashion.

If seeds of crops which beneficiaries are not familiar with are being distributed, simple planting and growing instructions in the local language, with clear diagrams, should be given out, to each family group.

There may be other tasks to be done, or some of the tasks may be divided up further. Remember that it is important to have women on the team, and to use women teams when distributing only to women. The last three tasks may require two or more people for each, depending on the size of the target group each day and the number of different items being handed out at one distribution. It is better to have too many staff present than too few, as the work is tiring and people will need to take rest-breaks. On average a team of seven people can distribute tools or seeds to 400 or 500 recipients in a day; more than this is not recommended. This is only possible when the distribution is well organised and there is local assistance with unloading the trucks.

Distribution site

The choice of site and how it is set up is very important. The site should have some shelter or shade. Buildings such as schools, clinics, and community centres are well-known to most people and can often be used for the day. Areas should be marked out with ropes or sticks on poles. Clear communication with the beneficiaries is essential. People should be told exactly where to wait, when and where to queue up, what information is required from them, and what to do next.

Monitoring

Once the seeds and tools have been distributed, monitoring can begin. The agreed indicators will need to be measured in order to determine the success of the project. For each project objective or activity there will be one or more measurable indicators, which will be such things as tools received, acres cultivated by each household, acres planted by each household, bags of grain harvested by each household, and drop in malnutrition rates. The monitoring team will need to:

- develop checklists of required information
- devise a simple recording system
- decide on a sampling method
- delegate responsibility for particular monitoring activities to team members
- inform local authorities about the monitoring exercise.

When the necessary information has been collected, the team should discuss and analyse the data and write up the results.

Monitoring offers another opportunity for close contact with some of the beneficiaries. It may be possible to gather more information than is required for measurement of specific indicators, and to observe the general post-distribution conditions and any changes. You may discover indicators that you had not thought of, or problems or benefits not originally anticipated. Monitoring is an opportunity to gather feedback and data useful for future development and modification of the project.

The beneficiaries should be involved in the process as much as possible. Discuss the results with them, and with others involved in the project, and ask for their response. Although monitoring has specific objectives it can be a very flexible process involving a range of people; sharing information and ideas and encouraging participation will help the rehabilitation of the community, and longer-term development.

Evaluation

An evaluation may have been planned for the end of the project or may be required before proceeding with a longer-term project or programme. Ideally, an evaluation should be done before the final report is written, so that findings and recommendations from the evaluation can be included. However, if the evaluation is done after the final report, the evaluator will have the advantage of being able to read it, and to interview the team after the final assessment process.

An independent evaluator should be contracted, from another part of the organisation or from outside. They will bring a fresh perspective to the situation, the inherent problems, the limitations and achievements, and will take a detached and objective view. The greatest benefit from an evaluation will be obtained when there is mutual co-operation, openness, and understanding between the evaluator and the project team.

Sometimes an evaluation will not have been planned, but may be essential because:

- a new approach or methods were tried
- things went wrong and the reasons why are not clear
- there seem to be important lessons to learn, which need to be more clearly defined
- approval or funding for a further project is conditional on an evaluation being carried out.

An evaluation may feel to the team like outside interference, and they may resent what they see as unjustified or unproductive criticism given out of context, with little real understanding of the actual situation in which the work was done. However, if the team is clear about what is needed from the evaluation, and they become involved in the process of setting out the objectives and terms of reference, then it can be a positive, productive experience of great benefit to all.

The people who received the seeds and tools, both present and future, are the ones who should benefit the most from any evaluation. Improving the efficiency and quality of assistance provided to vulnerable people is the main priority. It will be the responsibility of the project team to ensure that the evaluation includes close contact with the beneficiaries, local leaders, counterparts, and key informants. Feedback from women and men farmers or beneficiaries is crucial, as they are the ones who were directly affected. It is not uncommon for local officials and leaders to have different views from the beneficiaries about relief distributions.

Final project report and recommendations

The decision that comes at the end of the seeds and tools project is one that is often very difficult to make: should the project be extended or stop now? To make this decision requires careful analysis of information about current conditions. The checklists of questions in Chapter 3 could be used as a guide to re-assess the situation.

All the information from the base-line assessment, from project monitoring, plus the additional information your team has collected throughout the implementation phase, needs to be collated. A clear and detailed picture of the whole situation should emerge.

A team meeting to review the project could facilitate:

- pooling and consolidating all information to date
- analysing all data including feedback and lessons learned
- considering recommendations or requests from each community or group of beneficiaries for further support
- formulating recommendations for termination or continuation of the project.
- drafting the final project report.

If investigation shows that further assistance is not required the project must terminate, although this may be a difficult decision because of pressures from various sources for it to continue. A document giving an objective, systematic review and analysis of information, as the basis of a decision will be valuable.

Many seeds and tools projects do in fact continue for further growing seasons, because the original beneficiaries often require further assistance to re-build their livelihoods, or new groups of vulnerable people are identified. If the decision is to continue, the team is now back at the beginning of the project cycle (see Chapter 2). A team process as outlined in Chapter 5 will be required to develop an outline for future project recommendations; these can then be included in the final report. (See Appendix 9 for a recommended outline for a final project report.)

9 | Long-term considerations

Seeds and tools projects usually fit into the category of rehabilitation work. This work often comes after emergency interventions and before long-term development work. It therefore links the two and plays an important role in bridging any gap. In this final chapter, we look at seeds and tools distributions in a wider context of long-term development work. The aim must always be to build capacity and encourage self-sufficiency.

Rehabilitation work is designed to help people to move from extreme vulnerability towards relative stability and self-sufficiency. Seeds and tools project teams must be aware of the crucial linking function of the project from the beginning, and take responsibility for developing an approach that will support and strengthen local capacities, structures, and mechanisms. The team should maintain relationships of trust with the target communities and groups, which means keeping in close contact, seeking and responding to feedback, and using inclusive and consultative processes at every stage of the project: assessment, analysis, planning, implementation, monitoring, and evaluation. When the project comes to an end, all project information and resources should be handed over to the communities and organisations that will continue with the work.

Long-term agricultural work

When seeds and tools projects are continued as long-term agricultural programmes, it is often clear that additional areas of work will be essential to improve the sustainability of the livelihoods of farming communities. This assistance may take the form of agricultural extension work, involving training or skills transfer.

Even where refugees or internally displaced people are living in camps or settlements, newly acquired skills and knowledge can be very beneficial when their lives return to normal. Some areas of appropriate extension work are discussed below.

A general long-term aim of a rural development programme may be to improve local agricultural skills and knowledge by facilitating the sharing of relevant information from outside and within communities. Farming systems including soil management, improved crops, cropping systems, pest control, and marketing are the major areas of agriculture extension work. They are usually the priorities of government agriculture services for farmers. In some countries progress has been made during periods of stability only to be lost during times of crisis. Any so called progress that is too dependent on large centralised systems and infrastructures will be very susceptible to destruction during times of crisis.

It is not usually advisable to introduce new varieties of crops in a seeds and tools project. Outside support given during a longer-term project may help to facilitate some experimentation and crop trials. This combined with improved growing methods may slowly improve crop production at the local level.

Animal traction

Animal traction (ox ploughs and donkey carts) can be introduced or fostered to compensate for a lack of human labour, to improve productivity, and reduce work loads. It may require time and effort on the part of the project team to get the idea of animal traction accepted by the community. Qualified experienced trainers are essential. An animal traction project may involve providing equipment, animals, trainers, training facilities, and training for the manufacturing of equipment from local materials.

Farmer-to-farmer schemes

The best agricultural extension programme requires minimum support from outside and is carried out by the farmers themselves. A good example of this is the Campesino a Campesino or Farmer-to-farmer scheme in Nicaragua, Central America. Outside help was required at first to set up the project, but it soon became self-managed and self-sustaining. Farmers simply teach each other: groups of farmers visit one another's farms to share ideas, and there is a regular newsletter produced for members of the scheme. Good practices developed by experienced farmers are passed on so that they benefit everyone. The scheme is a great success and has been widely replicated.

Supporting marketing initiatives

Farmers' power in the market place can be increased through forming agricultural marketing collectives or co-operatives. These will be involved in storage, packaging transport and bulk sales. Quite often extension work can be carried out through these groups. Many co-operatives have been set up to facilitate the production of cash-crops by small-scale farmers. Government policies and agricultural departments tend to encourage these enterprises in order to stimulate national economic development.

While new hybrid varieties of cash-crops may bring benefit to better-off farmers who have sufficient resources to invest in large-scale production the small farmers may suffer. They lack bargaining power because they do not have the means to store and transport crops. They depend on others for these services and therefore must sell at any price offered to them at harvest time. By forming a group for bulk purchase of inputs, and joint marketing of produce, small farmers can greatly increase their incomes from cash-crops. Projects that support farmers' groups in ways that increase their input into agricultural development policies and their influence in the markets will help to improve the livelihoods of farming communities.

Rebuilding seed production systems

A seeds and tools project is necessary when farmers can no longer obtain sufficient quantities of reliable seed, for a variety of reasons. As well as supplying the necessary seeds as an immediate response, it may be possible for aid organisations to establish or re-establish seed systems that meet local needs in the longer term. Rebuilding sustainable livelihoods may depend on returning to old non-commercial varieties of crops. Small-scale producers, men and women, may be the best source of appropriate seeds. Such people may in the past have provided special, local varieties of seeds through local distribution or marketing systems. People are accustomed to dealing with these small-scale suppliers, and it may require very little additional outside support for them to re-establish their businesses.

When supporting local seed-producers is not possible, or only part of the solution, then a system involving larger seed producers may need to be developed. Farmers with the resources to grow seed in large enough quantities should be identified, and training provided if necessary. The inputs and time-span required are much greater in this case, and your organisation will have to be committed to a long-term project. Where government or commercial seed-production systems once existed, it may be possible to support them directly or indirectly. All that may be required is to provide seed to farmers until these sources resume their former scale of operations. It may be appropriate to hand over project resources to government departments; for example, pay-back schemes for seed could be managed by the Government Extension Department.

The technical skills and inputs required for large-scale seed production are usually too great for a small project to provide. Before making any commitment, an NGO should be certain that it is capable of organising the activity, and that such a seed-production system will benefit all farmers, including small-scale producers. It is clearly undesirable to help to re-establish a system and agricultural policies that were previously undermining community food security and small farmers' livelihoods, damaging the environment, and reducing local capacity to cope with crisis.

Seed banks

Local seed production and the use of seed banks will help to maintain genetic diversity. When local varieties have been virtually lost, a long-term project to re-introduce and multiply certain varieties will be required. Usually there will still be small amounts of seed of these varieties still obtainable, to start such a scheme. Or there may be alternative varieties that can be used temporarily until sufficient quantities of the required seeds are available.

It is vital to regain and maintain the wide selection of crop varieties formerly used by farmers. Farmers select seed to obtain varieties best suited to their conditions, and to do this a diverse gene pool or wide choice of varieties is essential. If germplasm or crop varieties have been lost from an area, international seed banks may be able to provide what is required in relatively small quantities; but to reintroduce these varieties will take time, skill, and commitment.

Supporting local tool production

Ensuring the availability of tools to all farmers and improving the quality of tools and equipment is another important area of long-term development work. After the initial emergency inputs, local production of tools should be encouraged. In some situations blacksmiths may need assistance to re-establish their trade; or a general shortage of blacksmiths may exist. In the latter case, community projects to train men and women in metal-working skills may be required. Existing blacksmiths could be involved in this from the beginning. This longer-term project may need to provide lost or improved equipment, training and introduce new methods to improve the quality of tools manufactured. A source of metal of good quality might be required by local blacksmiths because old supply systems have been interrupted, lost or are not adequate to meet present demands.

Building local capacity for emergency response

In situations where there is the possibility of disruption and instability it is essential to build local capacity to improve food security and cope with disaster. Projects that support and facilitate community-managed efforts to maintain agricultural services are most likely to survive a crisis. Skills and knowledge will not be lost, and local structures like farmers' groups, women's groups, distribution committees, co-ordination committees, seed banks, marketing co-ops, and micro-industries can quickly re-establish themselves, given a little support, during or after the crisis. These will provide valuable local services and play crucial roles in relief and rehabilitation work.

Appendix 1

Outline of an emergency preparedness plan

Introduction: present work and how this plan developed, evolved.

Background: history that led up to and including present situation.

Programme aim: and how these plans fit in with this aim.

Programme objectives: in summary or outline form with how these plans fit.

Programme strategies: in summary form with how these plans fit into each.

Scenarios: one for each potential situation describing each possible outcome. These are usually based on field assessments, project monitoring information, local knowledge and well informed situation analysis. Reference to these processes and relevant documents are important.

Activities: the planned interventions for each scenario above.

Resources: list resources that will be required, include list of potential or planned resources from all organisations if possible, list resources to be pre-positioned and resources pre-sourced to meet potential needs. In the case of seeds and tools pre-sourcing is most practical. Include project resources like vehicles, radios etc. That will be borrowed or need to be budgeted for and pre-sourced.

Management: how will responsibilities be met and by whom i.e. the anchor team; the assessment team/s; the manager; the field or distribution team/s; logistical, administrative, and programme support. List existing staff to be used and further staff requirements in each scenario.

Co-ordination: present and potential co-ordination with other organisations on the project plan and planned possible interventions.

Target groups: in each scenario and rationale behind targeting this group.

Distribution plans: for each scenario with a list of each kit or packaged provided to a group/family/individual.

Budget: showing the costs of seconded staff, borrowed project resources and/or used project resources as well as costs of all new project requirements for each scenario.

Appendix 2

Measurement conversion table

Metric with approximate British equivalents

1 centimetre = 0.394 inches
1 metre = 1.094 yards
1 kilometre = 0.6214 mile
1 square metre = 1.196 square yards
1 hectare = 2.471 acres
1 square kilometre = 0.386 square mile
1 cubic metre = 1.308 cubic yards
1 litre = 1.76 pints
1 decalitre (10 ltrs.) = 2.2 gallons
100 grams = 3.527 ounces
1 kilogram = 2.205 pounds
1 tonne (MT) = 0.984 ton

Temperature

To convert Fahrenheit into Centigrade:
subtract 32, multiply by 5, and divide by 9.

To convert Centigrade into Fahrenheit:
multiply by 9, divide by 5, and add 32.

Example: 20 degrees Centigrade or Celsius:
20 x 9 = 810 ; 180 /5 = 36 ; 36 + 32 = 68

So 20 degrees Centigrade equals 68 degrees Fahrenheit.

Appendix 3

Assessment report format

Note: These are guidelines only and may not be the format that your organisation uses.

Introduction: the reasons for carrying out the assessment and situation context.

Methodology: the assessment schedule, sampling techniques, team members and their roles,

The key informants: cross section of people and groups interviewed, difficulties met, difficulties with certain information, additional information gathered, when and where surveys were conducted, and number of surveys done and distribution by group.

Map: include a map or maps that show the region, area and location of all groups with important information from the surveys.

Findings: include information on each separate sector i.e. health, administration, agriculture.

- Include tables especially if certain frameworks were used in assessment and analysis.
- Point out conflicting information that is not reconcilable.

Possible evolution: an analysis of the situation/s and how they may progress if nothing is done.

Conclusions: the conclusions the team has come to about the situation.

Recommendations: what should be done to assist people in this situation. (This can include things that your organisation does not do as long as this is made clear. This may encourage other agencies or NGOs to take up this work.) Short-term and long-term work should be separated.

Appendix 4

Example of a mind map

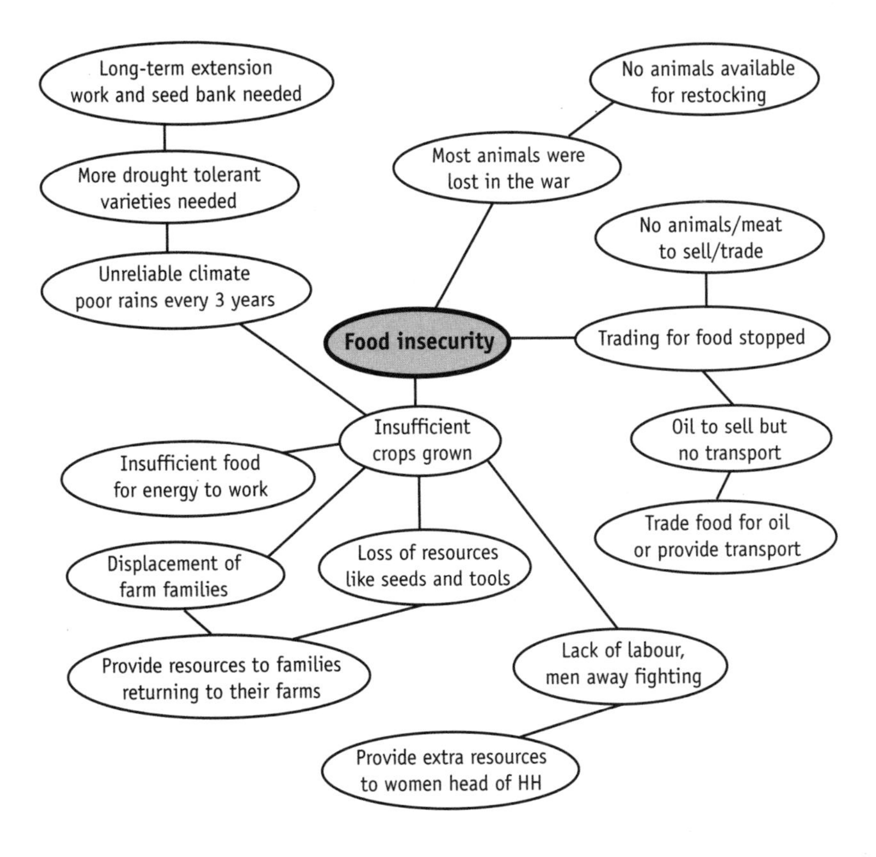

Appendix 5

Project budget

Staff costs

Salaries
Medical expenses
Allowances
Consultants' fees
Interview costs

International travel costs

Flights
Visas
Baggage costs

Accommodation & subsistence

House rent and services
Other accommodation
R&R
Camping equipment

Project administration

Office and warehouse rent, repairs and services
Phone and fax costs
Stationery and postal services

Vehicle capital costs

Vehicle purchase and duty
Accessories
Local travel/transport
Vehicle hire
Running costs (fuel, spares and insurance)
Maintenance and repairs
Workshop costs
Local flights

Capital equipment

Radios, faxes, satphones
Computers
Furniture

Relief goods

Seeds
Tools
Plastic sheeting
Bags, twine, tags

Local labour

Loading/unloading

Airfreight/shipping

Appendix 6

Botanical names of some common crops in English and Latin

African fan palm *Acacia albida*
African locust bean *Borassus aethiopicum*
Bitter leaf *Veronia amygdalina*
Black nightshade *Solanum nodiflorum and S. nigrum*
Calabash gourd *Artocarpus communis*
Bush greens, Amaranths *Amaranthus spp.*
Cabbage *Brassica oleracea*
Carrot *Daucus carota*
Cassava, manioc *Manihot utilissima*
Castor *Ricinus communis*
Cocoa *Theobroma cacoa*
Coconut *Cocos nucifera*
Cocoyam, Tannia, Yautia *Xanthosoma sagittifolium*
Common bean *Phaseolus vulgaris*
Cowpea *Vigna unguiculata*
Cucumber *Cucumis sp. and Cumumeropsis sp.*
Date palm *Phoenix dactylifera*
Eggplant *Solanum incanum*
Finger millet *Eleusine coracana*
Goa bean *Psophocarpus tetragonolobus*
Gourd *Cucurbita sp.*
Groundnut *Arachis hypogaea*
Lettuce *Lactuca sativa*
Maize *Zea mais*
Mexican sunflower *Tithonia*
Millet *Pennisetum americanum*
Native pear, Bush butter tree *Dacryodes edulis*
Oil palm *Elais guineensis*
Okra *Hibiscus esculentus*
Onion *Allium cepa*
Papaya, Pawpaw *Carica papaya*
Peas, chick peas *Cicer arietinum*
Earth pea, Bambarra groundnut *Voandzeia subterranea*

Pea *Pisum sativum*
Pigeon pea *Cajanus cajan*
Congo pea *Piper nigrum and P. guineense*
Potato *Solanum tuberosum*
Pyrethrum *Chrysanthemum cinerariaefolium*
Rice *Oriza sativa*
Safflower *Carthamus tinctorius*
Sesame, Beniseed *Sesamum indicum*
Shea butternut *Butyrospermum parkii*
Sorghum *Sorghum bicolor*
Soya bean *Glycine maximum*
Sugar cane *Saccharum officinarum*
Sunflower *Helianthus annuus*
Sweet potato *Ipomoea batatas*
Cocoyam *Colacasia antiquorum*
Tea *Camellia sinensis*
Tomato *Lycopersicon esculentum*
Turnip *Brassica rapa*
Wheat *Triticum aestivum*
Yam *Dioscorea spp.*

Quantity of proteins in some foods

cassava tubers – 0.9%
carrot, lettuce, tomato – 1.0%
mango,orange pawpaw – 1.0%
sugar cane – 1.0%
sweet potato – 1.1%
banana, okra fruit – 1.5%
tubers- taro, yam, potato – 2.0%
avacado – 2.0%
cocoyam-tubers and leaves – 2.5%
eggplant – 4.0%
black nighshade and bitter leaf – 5.5%
native pear fruit – 7.0%
cassava leaves – 7.5%
rice – 7.5%
shea butter fruit – 8.5%
maize – 9.0%
millet and sorghum – 10.0%
fonio, finger millet – 11.0%
wheat – 12.0%
earthpea pulses – 18.0%

seasame – 21.0%
pea and cowpea pulses – 22.0%
bean pulses – 24.0%
dried gourd and cucumber seeds – 24.0%
dried groundnut seeds – 32.0%
dreis soya beans – 40.0%

10.0% = 100mg of protein in 1000mg of food

Appendix 7

Calculating storage areas

Seeds

Grain seed is usually packed in 90kg or 50kg bags, and should be stacked at approximately one MT (10 × 90kg bags or 20 × 50kg bags) per pallet. (A pallet is a strong slatted wooden frame 1 square metre or approximately 3ft × 3ft in size.). Aisles between rows of pallets should be about 3ft wide for easy access. Ideally there should be one aisle per row, but one for every two rows is possible. This means that in a room 100ft × 100ft or 10,000 square feet capacity, you can store 550MT to 740MT.

Tools

Rows can be 9ft wide (the equivalent of three pallets) with aisles 3ft wide. For this spacing use a factor of 1.33 × the area of pallets to calculate the total area required, plus aisles. Ask the supplier for information about packaging, to help in planning storage area.

If the tools are without handles and are in bags then estimate the number of bags that can be safely stacked on a pallet (an area approximately 3ft × 3ft) and using the number in a bag calculate as follows:

[A] tools per bag × bags per pallet divided into the total number of tools

[B] × 9 square feet = the area in square feet that you will need for rows of tools;

[C] then, depending on the spacing you will use, multiply this area by 1.33 (or another factor) for total area with aisles.

Example:

[A] $$\frac{\text{10,000 hoes total}}{\text{25 hoes (per bag)} \times \text{16 bags} = \text{400 hoes per pallet}} = \textbf{25 pallets}$$

[B] 25 pallets × 9 square feet = **225 square feet**

[C] 225 square feet × 1.33 = **299.25 square feet total with aisles**

When tools are supplied with handles they are usually tied together in bundles; in this case substitute the number of bundles that can safely be stacked on a pallet and the number of tools in one bundle to get the number that fit on one pallet. Divide that into the total number of tools and then multiply by 9 square feet and multiply again by 1.33 to get total area with aisles.

When tools are packed in boxes then simply calculate the area in square feet which each box takes up (for example, 1ft × 1.5ft = 1.5 cubic feet). Then see how many can safely be stacked on top of each other, up to 6ft high.

Calculate as follows:

$$\frac{\text{total number of tools}}{\text{tools per box} \times \text{number of boxes high}} = \mathbf{A}$$

A × 1.5 square feet = total area for rows of tools;

then multiply by 1.33 for total with aisles.

This will give you space to make the rows of boxes of tools approximately 9 ft wide with 3 ft aisles between rows. If space is limited, wider rows will be necessary: use the factor of 1.25 instead of 1.33, which allows for 12ft rows and 3ft aisles.

Appendix 8

Sample distribution form

ORGANISATION NAME

Area:	Group/Community:
Date:	Distribution Site:
Head of Group/Community:	
Distribution Team:	Filled Out By:
Items Distributed:	

Head of Household	No. In Family Group	Registration No:	Area:	Items received:
Mary Smith	**5**	**A102**	**Kigola Village**	**1 kit**

Notes:

- The Head of Household or principal recipient will have to be determined according to the situation. It is best to have the criteria clearly defined to avoid problems. If they are the only family member who can receive the items distributed then this must be made very clear ahead of time.
- The number in a family group will depend on the definition of a family according to local customs or systems.
- The Registration Number can be optional but is useful when registration cards are used and checked on distribution day. Cards can be numbered in advance.
- Area, group, or leader can be used to help identify people in large mixed distributions.
- Items received can be listed with quantities or number and type of relief kits.

Sample registration card

DISTRIBUTION REGISTRATION CARD
FOR OXFAM LIBERIA SEEDS & TOOLS PROJECT

Area **Kigola Village**	Group
Leader	Date Registered
Name **Mary Smith**	No. In Family **5**
Registered by	Land Farmed (acres)
Registration No **A102**	

Distribution date:	items received
Distribution date:	items received
Distribution date:	items received
"	"

Appendix 9

Layout for final project report

Note: This is a guideline only, and may not be the same as the format used by your organisation.

Introduction

Background: brief history, geographic areas and assessments done

Problems identified: e.g. food insecurity

Brief summary: of aims, objectives activities and strategies

- Administrative set up
- Project implementation

Activities: what activities have been carried out, when, where and with who

Beneficiaries: how did the target group/s benefit and numbers of beneficiaries

Distributions: methods, relief and other resources and problems encountered.

(Refer to tables/reports of distributions carried out.)

Objectives: to what degree have the project objectives been met using indicators to show progress, achievements and short falls.

Strategies: to what degree have the strategies worked in reaching objectives.

Changes: what changes took place during the project cycle.

Impact: what impacts did the project have on the lives of the beneficiaries both positive and negative, planned or unplanned.

Capacity building: did the project help to build local capacities and/or strengthen partner agencies?

Co-ordination/co-operation: with other agencies, ngos

Important lessons: lessons learned from the project in any area.

Achievements: was the project successful and viable, against agreed indicators.

(Results from an assessment or evaluation including input from the beneficiaries can be used.)

Budget performance: the budget with any changes or modifications explained.

Further reading

Appert J (1988) *The Storage of Food Grains and Seeds*, Macmillan

Carruthers I and Rodriguez M (1992) *Tools for Agriculture: A Buyer's Guide to Appropriate Equipment for Smallholder Farmers*, ITP.

Casley D and Kumar K (1988) *The Collection, Analysis and Use of Monitoring and Evaluation Data*, Johns Hopkins University Press.

Cooper D , Vellve R and Hobbelink H (1992) *Growing Diversity: Genetic Resources and Local Food Security*, ITP.

Cromwell E, Wiggins S and Wentzel S (1994) *Sowing Beyond the State: NGOs and Seed Supply in Development Countries*, Westview Press.

Dupriez H and de Leener P (1992) *African Gardens and Orchards: Growing Vegetables and Fruit*, Macmillan.

Feldstein H S and Jiggins J (1994) *Tools for the Field: A Methodologies Handbook for Gender Analysis in Agriculture*, ITP.

Feuerstein M-T (1987) *Partners in Evaluation: Evaluating Development and Community Programmes*, Macmillan.

Gosling L and Edwards M (1995) *Toolkits: A Practical Guide to Assessment, Monitoring, Review and Evaluation*, SCF.

Harries D and Heer B (1993) *Basic Blacksmithing: An Introduction to Toolmaking*, ITP.

Hubbard M (1995) *Improving Food Security: A Guide for Rural Development Managers*, ITP.

Keen D (1993) *Famine, Needs Assessment, and Survival Strategies in Africa*, Oxfam.

MacDonald I and Low J (1984) *Tropical Field Crops*, Evans.

Macrae J and Zwi A (eds) (1994) *War and Hunger: Rethinking International Responses to Complex Emergencies*, Zed Books with SCF.

McRobie G (ed) (1989) *Tools for Organic Farming*, ITP.

Mikkelsen B (1995) *Methods for Development Work and Research: A Guide for Practitioners*, Sage Publications.

Miller G and Dingwall R (eds) (1997) *Context and Method in Qualitative Research*, Sage Publications.

Moris J and Copestake J (1993) *Qualitative Enquiry for Rural Development*, ITP.

Oakley P (1991) *Projects with People: The Practice of Participation in Rural Development*, ILO.

Okali C, Sumberg J and Farrington J (1994) *Farmer Participatory Research: Rhetoric and Reality*, ITP.

Poston D (1994) *The Blacksmith and the Farmer: Rural Manufacturing in Sub-Saharan Africa*, ITP.

Pratt B and Loizos P(1992) *Choosing Research Methods: Data Collection for Development Workers*, Oxfam.

Rappaport R (1992) *Controlling Crop Pests and Diseases*, Macmillan.

Rubin F (1995) *A Basic Guide to Evaluation for Development Workers*, Oxfam.

Slocum R, Wichhart L, Rocheleau D, and Thomas-Slayter B (eds) (1995) *Power, Process and Participation: Tools for Change*, ITP.

Smyth I, and March C (forthcoming) *A Guide to gender-Analysis Frameworks*, Oxfam.

van Veldhuizen L, Water-Bayer A and de Zeeuw H (1997) *Developing Technology with Farmers: A Trainer's Guide for Participatory Learning*, ETC Netherlands.

Walker, B (ed) (1994) *Women and Emergencies*, Oxfam.

Williams C N, Uzo J and Peregrine W T H (1991) *Vegetable Production in the Tropics*, Longman.

Young H (1992) *Food Scarcity and Famine: Assessment and Response*, Oxfam.